Organic Chemistry
A Guided Inquiry
for Recitation, Volume 2

---

a process oriented guided inquiry learning course

POGIL

Andrei Straumanis

BROOKS/COLE
CENGAGE Learning

Australia • Brazil • Japan • Korea • Mexico • Singapore • Spain • United Kingdom • United States

Organic Chemistry: A Guided Inquiry, Volume 2
Andrei Straumanis

Publisher: Mary Finch

Acquisitions Editor: Chris Simpson

Editorial Assistant: Laura Bowen

Marketing Manager: Barb Bartoszek

Marketing Assistant: Julie Stefani

Project Manager, Editorial Production:
  Michelle Clark

Art Director: John Walker

Print Buyer: Linda Hsu

Permissions Editor: Dean Dauphinais

Manufacturing Manager: Marcia Locke

Production Service: PreMediaGlobal

Content Project Management: PreMediaGloba

For product information and technology assistance, contact us at
Cengage Learning Customer & Sales Support, 1-800-354-9706

For permission to use material from this text or product,
submit all requests online at **www.cengage.com/permissions**
Further permissions questions can be emailed to
permissionrequest@cengage.com

ISBN-13: 978-1-111-57398-0

ISBN-10: 1-111-57398-0

Brooks/Cole
20 Davis Drive
Belmont, CA 94002-3098
USA

Cengage Learning is a leading provider of customized learning solutions with office locations around the globe, including Singapore, the United Kingdom, Australia, Mexico, Brazil, and Japan. Locate your local office at **www.cengage.com/global**

Cengage Learning products are represented in Canada by Nelson Education, Ltd.

To learn more about Brooks/Cole, visit **www.cengage.com/brookscole**

Purchase any of our products at your local college store or at our preferred online store **www.cengagebrain.com**

Printed in the United States of America
2 3 4 5 6 7 19 18 17 16 15

This book is dedicated to my friend and colleague, Rick Moog, and to his tireless, selfless work in support of student learning.

Acknowledgements

Huge thanks go to the organic chemistry students at the University of Washington and College of Charleston. In preparing these ChemActivities nothing could have substituted for watching real students work in a real classroom setting. Thanks also to my colleagues who have been generous with their knowledge of chemistry and chemistry students.

Particular thanks go to chairs in both my departments: Jim Deavor at the College of Charleston, and Paul Hopkins and Philip Reid at the University of Washington. Their support of this project is yet further proof of their dedication to chemistry instruction, and their willingness to take risks to keep their departments on the cutting edge. It is not a coincidence that under their leadership both departments sit atop their respective institutional categories in the key metric of number of chemistry majors.

Many of the over 100 faculty who used the full version of this workbook (Organic Chemistry: A Guided Inquiry, 2e) have contributed to improvements in this recitation edition. Thanks for all of your suggestions and corrections.

Thanks to the POGIL Project and to its growing number of participants. Special thanks to Rick Moog, whose contagious enthusiasm for guided inquiry inspired me and many others to embark on this path.

This work would not have been possible without the POGIL Project and its primary funder, the National Science Foundation. The United States Department of Education's Fund for Improvement of Post-secondary Education (FIPSE) supported my work with very large classes at the University of Washington under grant number P116B060026.

Words cannot express the depth of my love and appreciation for my wife, Allyson Chambers, MD, and for my three sons, Milo, Luca, and Nico, purveyors of the thunder of tennis shoes that rolls through my house each afternoon.

Comments from Faculty

'Organic Chemistry: A Guided Inquiry' was a true revelation to me. In adopting a POGIL format in a large classroom my day-to-day preparations were comparable in intensity and duration to the time I spent preparing for traditional lectures. In my years as a college educator, I have not seen anything as pedagogically powerful as a POGIL class using Straumanis' workbook. I believe the future of college 'teaching' lies in this type of 'learning.' I give Straumanis my highest rating.

Dr. Stefan Kraft, Kansas State University

This workbook has revolutionized the way I teach organic chemistry. The students process the material in logical steps, are active learners in the classroom, and the end result is a deeper understanding of organic compounds and reactions. I highly recommend this book!

Dr. Bruce J. Heyen, Tabor College

The guided inquiry helps me [the professor] think more like a student and it helps me cover material more efficiently.

Dr. Dan Esterline, Heidelberg College

Organic Chemistry: A Guided Inquiry is a great way to teach and learn organic chemistry. The students love the interactive nature of class time. I become better acquainted with each of my students, which enables me to tailor my teaching to maximize each student's learning. It has transformed the way I teach.

Dr. Timothy M. Dore, University of Georgia

It is fabulous to see the increase in understanding of reaction mechanisms. The students learn the mechanisms by working together, discussing, and sometimes arguing about them, but they don't memorize them. What I really like about the approach is that when class is ending, the students are still working. There is no clock watching!

Dr. William Wallace, Barton College

I am thankful that I decided to use the Guided Inquiry approach in my class. Students have responded in a very positive manner. Other faculty, too, see a change in the students' attitude towards learning Organic Chemistry.

Dr. Karen Glover, Clarke College

I have been using the Guided Inquiry Organic Chemistry materials since they were in manuscript form, and I am delighted about the second edition. The ChemActivities do an excellent job enabling students to build on their knowledge to develop new knowledge and to apply chemical concepts to new situations. Students enjoy organic chemistry and they learn it well because they engage directly with the material and with their peers.

Dr. Laura Parmentier, Department Chair, Beloit College

Watching my students engaged in discussing organic chemistry in their groups during organic class has convinced me that I made the right decision to change to POGIL after twenty years of brilliant lecturing. Straumanis' ChemActivities really do help the students learn organic much better than my lectures ever did.

Dr. Barbara Murray, University of Redlands

Comments from Students

I didn't get tired during class because I was constantly thinking and working instead of in a lecture class where I just listen and get easily tired.

The act of explaining the concept forced me to clarify the concept in my own head.

Class time was actually learning time, not just directly-from-ear-to-paper-and-bypass-brain-writing-down time. Learning the material over the whole term is far easier than not "really" learning it until studying for the tests.

I wasn't just blindly copying notes on the board but actually working through problems and learning.

Overall, I was far less stressed than many of my friends who took the lecture class. They basically struggled through everything on their own.

Group work has helped me find motivation for studying.

It was hugely beneficial to be able to discuss through ideas as we were learning them; this way it was easy to immediately identify problem areas and work them out before going on.

The method of having us work through the material for ourselves— as opposed to being told the information and trying to absorb it—makes it seem natural or intuitive. This makes it very nice for learning new material because then we can reason it out from what we already know.

I felt like I was actually learning the information as I received it, not just filing it for later use. The format helped me retain much more material than I have ever been able to in a lecture class, and the small, group atmosphere allowed me to feel much more comfortable asking questions of both other students and the professor.

Through working in groups it was nice to see where everyone else was in understanding the material (i.e. to know where other people were having trouble too).

We were able to discover how things happened and why for ourselves…instead of being told.

Advice from Students to Students

Don't let yourself take the course lightly just because class is fun and relaxed (and goes by fast!). Do the homework and reading.

Give yourself some time to settle into group learning. Lots of us did not think we would like it or that it would work. It does.

Don't fall behind. Playing catch up is not fun. Don't be afraid to ask questions and argue in your group. That is the way learning is done in this class.

Find a study group ASAP and meet regularly [outside of class] every week. I wish I had done this sooner.

Overview of the POGIL Method

Process Oriented Guided Inquiry Learning (POGIL) is used to teach tens of thousands of science students each semester, and to date, over 150 colleges and universities have used POGIL to teach organic chemistry.

These materials are flexible and can be used in a variety of settings, but during most POGIL sessions...

- Students work in teams of three or four to answer carefully designed **Construct Your Understanding Questions**, which guide student groups toward discovery of a chemical concept.

- The teaching assistant, peer leader, or instructor serves as the facilitator of learning, not as the primary source of information. Though effective facilitators spend much of class time observing student group work, they do not sit on the sidelines. During each class they ask and answer questions, lead whole-class discussions, and deliver just-in-time mini-lectures.

For more information, read the chapter in this book entitled Introduction to POGIL, visit www.pogil.org/straumanis, or contact the POGIL Project at www.POGIL.org.

Another good resource is the instructor-only Yahoo group moderated by the author of these materials, and dedicated to POGIL Organic Chemistry (find it by searching "**giorganic**" at groups.yahoo.com).

Forward on Process Oriented Guided Inquiry Learning (POGIL)

by Rick Moog, Jim Spencer and John Farrell, authors of *Chemistry: A Guided Inquiry* (for General Chemistry) and two volumes of *Physical Chemistry: A Guided Inquiry*.

These guided activities were written because much research has shown that more learning takes place when the student is actively engaged and when ideas and concepts are developed by the student, rather than being presented by an *authority*—a textbook or an instructor. The *ChemActivities* presented here are structured so that information is presented to the reader in some form (an equation, a table, figures, written prose, etc.) followed by a series of questions that lead the student to the development of a particular concept or idea. Learning follows the scientific process as much as possible throughout. Students are often asked to make predictions based on the model that has been developed up to that point, and then further data or information is provided that can be compared to the prediction. In this way, students simultaneously learn course content and key process skills that constitute scientific thought and exploration

How to Use This Book

To Instructors

This book was developed for faculty who want to take advantage of the benefits of Process Oriented Guided Inquiry Learning (POGIL) during a supplemental class. Consistent with POGIL's focus on skills and team building, **research indicates POGIL increases a student's success in a subsequent lecture**.[1] The proposed mechanism is that the learning skills, confidence, and study group behavior developed in a POGIL environment help students get more out of lecture, readings, homework, etc.

The activities in this book are designed to be a student's first introduction to a topic. This means your TAs will be *ahead* of you, and your lecture on a topic should come *after* students have encountered that topic in recitation. This may be the opposite of what you are used to, but I think you will find that giving students a chance to discover a topic via a POGIL ChemActivity is superior to preparing for lecture by reading (or *not* reading) a traditional textbook chapter. I have found that a POGIL introduction to a topic greatly increases students' appreciation of lectures. You may even find that you can spend less time working through basic examples, or move more quickly through topics that gave your students trouble in the past.

This workbook is designed to complement any textbook. Most courses will not use all the activities herein. Choose the activities you like, and use them in whatever order works best for your course.

The once-per-week, student-friendly activities are designed for supplemental classes, but can also be used in lab, for homework, or as the basis for a hybrid POGIL-lecture approach.

The first chapter (**Introduction to POGIL**, starting on page 1) serves as a brief training for teaching assistants, undergraduate peer leaders, or instructors. POGIL is a *learning* method, so both students and instructors need to be familiar with POGIL, and will benefit from reading this chapter.

To Students

This book is designed to make organic chemistry more enjoyable and less intimidating, but without sacrificing depth. Too many organic chemistry students memorize facts, only to forget them after the exam. This workbook guides you toward a deep understand so you learn more, retain it longer, and do better in subsequent courses and on standardized exams.

This book is designed to be used during class. For each activity, read the **Model** then work with your group to answer the **Construct Your Understanding Questions** that follow. Stay together! For each question, it is a good idea to compare answers within the group before moving onto the next question.

If you are unsure of an answer after checking with your group, some good strategies are to read the next question, ask a nearby group, or pose a question to the TA or instructor, but instead of asking "Is our answer right?" ask a question that helps the instructor understand the source of your confusion.

It is your collective responsibility to manage your time so that you finish these questions and even get to some of the **Extend Your Understanding Questions** before the end of class. Before the next class, finish the *entire* activity including the **Confirm Your Understanding Questions**.

Be sure to read the *Advice from Students to Students* section on page vi, and the Introduction to POGIL (starting on page 1) so you can make the most of your efforts in this course. Many students find it useful to read these sections at the start of the course, and again two weeks into the course.

---

[1] 1) Moog, R.S.; Spencer, J.N., eds., American Chemical Society Symposium Series, American Chemical Society, Washington, DC, 2008. 2) Straumanis, *unpublished results*.

Contents

Notes

Introduction to POGIL

Permission to use material in this chapter has been generously provided by the POGIL Project.

Key Issues

This first page is a list of key considerations regarding a use of the POGIL method. The second page is a chronological account of a typical POGIL session. Both are provided in recognition of busy schedules, though students and instructors will likely benefit from reading this entire chapter.

- Each activity in this book is designed to be the introduction to a topic. This means each topic covered using POGIL should appear in recitation *before* it is covered in lecture.

- Start each POGIL session with a one-question **quiz** covering material from the *previous* class.

- Basic format: instructor-assigned groups of 3 or 4 complete a ChemActivity by reading the **Model** and working together to answer the **Construct Your Understanding Questions**.

- On the first day of class, many student groups are shy about talking. It helps to designate one person as **manager** and have this person read each question aloud or, at the very least, ask if everyone is in agreement before moving onto the next question. Manager duties should rotate.

- Students who participate in their group during class are far more successful on quizzes and exams than students who work without interacting with their group.

- Many students have had bad experiences with group grading. It is therefore worth emphasizing that in this course, **group work does not mean group grading**! The purpose of group work is to help each student prepare for quizzes and exams, taken *individually*.

- The instructor should not spend more than a few minutes explaining how POGIL works. Students figure it out quickly, and can read this chapter for more information.

- POGIL instruction is three parts listening and one part talking. Spend the time to figure out what is causing a group's confusion and you may only need a few sentences to clear it up.

- The instructor should not sit on the sidelines and expect the materials to work without help. When not answering a question, rotate through the class listening to each group, monitoring progress, especially on key questions (marked with a key), and intervening when necessary. Get to know the students so you can guide and inspire them to exceed their own expectations.

- Reading on a topic in the regular textbook should follow (not come before) a POGIL activity.

- Student groups should complete as many questions as possible during class, and may choose to meet after class, but each student must finish the *entire* activity before the next class.

- Students who feel rushed during a POGIL session should preview the activity by reading the models and questions *before* class.

- Students: Asking the instructor the question "Is our answer right?" is not as useful as asking questions like: "We are concerned our answer is wrong because…"

- Students are *not* provided an answer key to the in-class Construct Your Understanding Questions because doing so short-circuits *the* critical step in the learning process. Learning is largely the act of confirming your answers via discussion with your group, other groups, or the instructor; doing homework problems (for which there *is* a key); and reading in your textbook.

- Instructors: When faced with the question, "Is our answer right?" be supportive but firm. Find out if the group is in agreement, or refer them to the relevant part of the Model. Ask students to explain their reasoning or rate their confidence level. If students are frustrated, stay and help them talk it out or promise to return and check that they achieved closure.

Account of a Typical Day in a POGIL Recitation Section

The following is a description of a typical day in a POGIL class of about 25 students. Most items on the list below are described in greater detail on the pages that follow.

- Seating chart with names (and roles) is posted on-line and on the door.

- On the first day of class, bring copies of the activity for students who do not bring a book.

- Each group has a folder (numbered or color coded). The first person in a group to arrive picks up the group folder, which contains graded work from the previous class. As each group member arrives, she retrieves her graded quiz from the folder.

- The group folder also contains blank copies of the upcoming quiz (face down in the folder). When the instructor gives the signal to begin the quiz, the Manager distributes a copy of the quiz to each group member. (Quizzes are taken individually.)

- After three minutes, instructor calls time and the completed quizzes are turned in or placed in the group folder.

- Instructor briefly goes over the quiz. (Alternatively, a longer period of time can be devoted to more open-ended group discussion of the quiz.)

- Instructor briefly (< 3 minutes) reviews the key concepts from the previous class, and very briefly (< 1 minute) previews the upcoming activity.

- Students are instructed to begin work on the new activity. It is best to write on the board how many minutes groups have to work on a given set of questions. (For example: "You have 5 minutes to work on Questions 1-3." This tells students that Questions 1-3 are fairly easy or review, and helps them manage their time effectively.

- The instructor circulates to each group and briefly checks that work has begun.

- On a second pass, the instructor takes more time to listen to groups, assess progress, and examine answers to the Construct Your Understanding Questions, intervening if necessary. Many times, a group will find their own errors, so not all errors require intervention.

- The instructor responds to questions when the manager of a group raises her hand. (Before responding, make sure the group has already discussed this question *within* the group.)

- At the end of the time allotted for the first group of questions (e.g. Question 1-3) the instructor pauses group work, and asks one or more groups to report an answer to a key question. If necessary, the instructor can ask for student commentary on this answer.

- The directive on the board is replaced with a new directive (e.g. "You have 20 minutes to work on Questions 4-8.") These directives can be compiled into a PowerPoint or overhead presentation that might also include pre-prepared examples or auxiliary questions.

- One or more times during class, but always at the end of class, the instructor interrupts group work for a mini-lecture or whole-class discussion. This can consist of a group's Presenter reporting an answer followed by instructor-led discussion of that answer. This type of summary is especially necessary when students are having difficulty with a topic.

- If you have time, class can end with preparation of a recorder's report. This report may be focused on content (e.g. "Write down the three most important concepts you learned today, and any questions that remain unanswered.") or process ("What is one strength and one area for improvement for your group's performance today?").

- The class ends on time.

- Student groups often choose to stay after class to complete the activity. If the classroom is occupied during the following class period, the instructor may want to suggest or arrange for a space for continued work (e.g. a nearby classroom or in the hall outside class).

How People Learn

Research on learning tells us that students learn best when they are…

- actively engaged and thinking in class.[1]
- given an opportunity to construct their own understanding.[2,3]
- discussing ideas and asking questions as part of a dynamic and social team.[4,5]

The research that inspired POGIL is not new, and POGIL is not the first classroom method that has put this research into practice. If you have been working to make your classroom more active or participated in group learning, it is likely you will find some POGIL techniques familiar. Yet it is best to suspend any doubts, preconceptions, or fears about organic chemistry or group learning. They will only interfere with your success and enjoyment of this class. At the end of a POGIL class, less than 10% of students are negative about POGIL,[2] but many students take several weeks to come to this conclusion. To get the most out of this course, read this chapter once before the start of class, and then again a few weeks into the course. For further reading, you may wish to consult the following excellent resources on teaching and learning.[1-10]

What is POGIL?

A POGIL classroom differs dramatically from a traditional classroom in that there is little formal lecture. The instructor serves as the facilitator of learning rather than the primary source of information and students work in self-managed teams to analyze data and draw conclusions, modeling the way a team of scientists function in the research laboratory.

POGIL's central claim is that it helps students simultaneously develop content knowledge and key process skills. The hypothesized mechanism is two-fold, stemming from marriage of the PO (Process Oriented) and GI (Guided Inquiry) methodologies implied in the name. The "GI" part is achieved via use of carefully designed learning cycle[6,7] activities that guide students toward construction of their own understanding. Such discovery experiences have been shown to improve confidence while helping students to understand and remember more.[4,5]

The "PO" part comes from the frequent or exclusive use of small groups. There is a large body of evidence suggesting that cooperative learning fosters positive attitudes toward the subject matter, as well as growth in process skills such as critical thinking, teamwork, and metacognition.[1] Perhaps most importantly, the positive interdependence generated in a small group setting has been shown to attenuate the feelings of isolation, disorientation and competition that often correlate with underachievement or failure in a traditional classroom environment, especially for women and minorities.[8,9,10]

---

[1] Bransford, J.D.; Brown, A.L.; Cocking, R.R., *How People Learn*; National Academy Press: Washington DC, 1999.

[2] Abraham, M. R., Inquiry and the learning cycle approach. In Chemists' guide to effective teaching (Vol. 1), ed.; Pienta, N.J.; Cooper, M.M.; Greenbowe, T.J., eds. Prentice Hall: Upple Saddle River, NJ, 2005.

[3] Lawson, A. E., "What Should Students Learn About the Nature of Science and How Should We Teach It?" Journal of College Science Teaching 1999, 401-411.

[4] Bruffee, Kenneth A. *Collaborative Learning: Higher Education, Interdependence, and the Authority of Knowledge.* Baltimore: Johns Hopkins Press, 1993.

[5] Johnson, D. W.; Johnson, R. T; Smith, K. A. *Active Learning: Cooperation in the College Classroom*; Interaction: Edina, MN, 1991.

[6] Karplus and Thier. *A New Look at Elementary School Science.* Chicago, Rand McNally (1967).

[7] Piaget, J.; *J. Res. Sci. Teach.* 1964, 2, 176.

[8] Hewitt, N.A.; E. Seymour, *Factors Contributing to High Attrition Rates Among Science, Mathematics, and Engineering Undergraduate Majors: A Report to the Sloan Foundation.* 1991, Denver: U. of Colorado Press.

One Explanation for Why POGIL Works (the "Aha!" moment)

"Aha! I get it!" The power of the POGIL method is manifest in the frequency with which such "Aha!" moments take place during class. Consider a student who at first struggles with a concept and becomes frustrated, but–through a combination of answering guiding questions, analyzing data, discussion with group mates, and (as needed) instructor facilitation–the concept finally "clicks" for the student. The culmination of this dissonance-resolution process is often marked by an utterance: "Aha!" or "I get it!" Flushed with the excitement of discovery, this student becomes an advocate for the concept. As if she were the first person to ever discover it, the student works to help her group also appreciate it. In explaining the concept to them she deepens her own understanding, perhaps developing a question for the facilitator about its implications, limitations, or connections to other material.

Such cognitively rich experiences (discovery, discussion, learning by teaching, asking questions, etc.) are established pathways to conceptual understanding and long-term retention, and help explain POGIL's effectiveness.[1] However, the most lasting outcomes of such an experience may be related not to content, but to skills, particularly metacognitive skills (learning how to learn). The learning experiences in a POGIL classroom enhance a student's experience of lecture, homework, and other course offerings because it awakens students to the idea that learning is not memorization of facts and rules (boring), but a creative process requiring thought, participation, and discussion (fun). The next time this student encounters frustration in her studies, she is less likely to give up, isolate herself, or resort to memorization. Knowing what it feels like to figure out and eventually own a concept will motivate her to use the resources at her disposal, especially discussion with others. The pathway to understanding provided by a POGIL learning environment appears to rewire many young learners' natural fear of dissonance—and transform it into the hallmark of a lifelong learner: the confidence that *not* understanding is just an exciting (though sometimes frustrating) prelude to the satisfaction of understanding something new.

The Structure of a POGIL Activity:

The learning cycle, developed by Karplus and Piaget,[6,7] is similar to the scientific method.

We have found that activities that follow a learning cycle are very effective at generating Ah ha! moments, and encouraging development of targeted process skills such as critical thinking and self-assessment.

The Learning Cycle

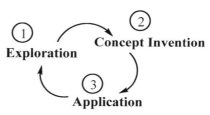

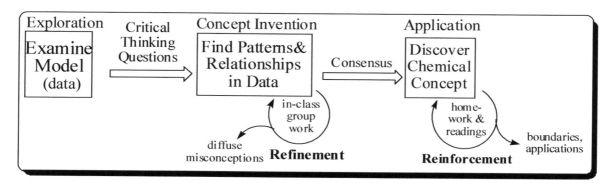

[9] Tobias, S., "Women in Science - Women and Science" *J. of Col. Science Teach.* 1992, 21, 276.
[10] Treisman, P.U. Innovations in Educating Minority Students in Math and Science; Dana Foundation: 1988.

Exploring the Model: Directed Questions

A POGIL learning cycle activity begins with a Model. This model often contains enough information such that a group of students could extract from it the target concept. However, to help students, Construct Your Understanding Questions are provided to guide students toward this concept. The questions usually begin with the very simple. Such questions help with the Exploration phase, and may simply direct students to look at the appropriate part of the model (for this reason such questions are called directed questions). For example, a directed question for the figure below might be: "What aspect of POGIL implementation helps diffuse student misconceptions?"

Coming to Consensus: Convergent Questions

After a few directed questions, the activity usually contains a question that requires students to process the data and find patterns. This is not a clean, linear process, but good questions will bring most student groups to a consensus that resembles the targeted concept. These types of questions are termed convergent questions since most student groups will converge on the same answer.

Though many student groups will converge on the same answer to a convergent question, they still may lack confidence in their answer. Confirmation by the instructor is often not the most valuable next step. As described below, it can be much more useful for the instructor to allow groups to confirm or correct their own answers via the application phase.

Confirming Your Understanding: Application Questions

The final stage of the learning cycle is Application. The purpose of application questions in a successful POGIL activity is to help students assess their current understanding, then assimilate this student-level understanding with expert explanations, including familiarizing themselves with expert terminology. This is best done with a combination of group discussion, teacher talk (e.g. a summary mini-lecture), homework, and reading from a complementary textbook. Textbooks provide a concise explanation of key concepts in expert terminology, but most students are not ready to read this explanation until after they have had a chance to discover the concepts and generate explanations in their own words. For this reason, most POGIL courses ask students to read the assigned sections of the text *after* they have completed the activity on that topic.

In general, forcing students to figure out if they have the concepts is far superior to simply publishing the answers to the in-class Construct Your Understanding Questions. Any time students work toward their own understanding they will learn more and retain it longer. When students are given an answer key, there is a great temptation to look at the expert answers without fully constructing their own answers, and this circumvents the critical part of the learning process.

Content and Process Skill Goals

Most instructors have clearly defined content goals for their course. Their syllabi might list topics like "Nucleophilc Substitution" or "Acid-Base Reactions." A teacher's content goals are often tied to subsequent courses, for which their course may be a prerequisite.

A course can also be designed with specific process skill goals in mind. Development of the key process skills listed at right will help students learn course content, and create new knowledge, applicable to other courses and contexts.

POGIL Process Skills[11]
Information Processing
Critical Thinking
Problem Solving
Communication
Teamwork
Management
Self-assessment

---

[11] Moog, R. S.; Creegan, F. J.; Hanson, D. M.; Spencer, J. N.; Straumanis, A., Process Oriented Guided Inquiry Learning. In *Chemists' guide to effective teaching (volume 2)*; Pienta, N.J.; Cooper, M.M.; Greenbowe, T.J., eds. Prentice Hall: Upper Saddle River, NJ, 2009.

Is my answer the *correct* answer?

Perhaps the most important but least self-explanatory skill on the list above is self-assessment: the ability to asses if you have understood a concept. A student does not need self-assessment skills to memorize facts and algorithms, to churn out answers to familiar exercises, or to check his answers on an answer key; so many students are not familiar with this skill.

Of course, on the exam *and in life*, there is no answer key. Many students are at first resistant to the frustration inherent in practicing self-assessment. This is the frustration of not knowing if your answer is the right answer. Experiencing and ultimately overcoming this frustration is a key part of becoming a mature learner.

A common student complaint in a POGIL classroom is the following: "How do we know if we are learning the right things if you don't tell us the right answers?" Advice on answering this question is given in the FAQ entry at the end of this chapter entitled **How do you answer the question: "How do I know if my answers are correct?"** One answer is that during an exam it is useful to know if you are right or wrong.

Experienced students sometimes describe self-assessment as the feeling you get when you finally figure something out...the "Ah ha! moment." Or, conversely, the ability to recognize the absence of that feeling: the confusion and frustration associated with not understanding *yet*. **This may be the central skill of all science.** Without this skill a student does not know when to keep asking questions and when his preparations for the exam are complete, and a researcher does not know when to do more experiments and when to publish her findings.

Assigning Group Membership

At the beginning of a semester start with alphabetical group assignments. Then, for the first few weeks, change weekly, mixing randomly so that students get to know a variety of other students and learning styles. Many instructors strive, by mid-course, to have students in a set of static and functional groups so they can begin building specific dynamics. Others continue changing groups throughout the course.

There are three basic strategies for assigning groups:
* Instructor assigns heterogeneous groups based on quiz or exam scores
* Students are allowed to self-select into groups (usually resulting in homogeneity)
* Instructor assigns groups organically based on a variety of criteria (e.g. functionality)

Instructor assignment of group membership is an opportunity to gain valuable insight into the personalities of your students. The more you know about your students, the better you will be able to place them in functional groups. For even a class of twenty this can be a time consuming but rewarding process.

When assigning group membership based on skills and personalities, a useful trait to think about is assertiveness. (This may be more important than gender or other variables.) In general, do not put one quiet, unassertive person in a group with three loud, pushy people. Some of the most successful groups result from putting four quiet, thoughtful students together, or four loud, talky students together.

Formal Roles

One of the great challenges of group work is achieving equal participation among group members. Equal participation is important both in terms of content goals (students who do not participate may not learn as much), and process goals (students who do not participate will not have an opportunity to develop key process skills).

Many POGIL classrooms employ the formal roles described below. Often roles are rotated every class meeting so that every student experiences each role and its responsibilities. Some instructors have had success rotating roles less frequently. It can be helpful to distribute color coded cards, pins, or lanyards, each with the name and description of a role. You can also simply designate that the person with the group folder (see section on Course Management) is the manager, the person to the right of the manager is the presenter, etc..

Even without formal roles, an instructor can informally encourage equal participation. For example, a question may be posed directly to a less active student. Similarly, the instructor can call on certain individuals to present information at the board, or to serve as spokesperson for the group. The following are commonly used roles.[12]

Manager   Manages the group. Ensures that members are working together, no one is left behind, and that assigned tasks are being accomplished on time, including that all members of the group participate in activities and understand the concepts. The instructor responds only to questions from the manager who must raise his or her hand to be recognized. This encourages groups to do some internal processing of a question instead of immediately asking the instructor.

Presenter   Presents oral reports to the class. These reports should be as concise as possible; the instructor will normally set a time limit.

Recorder   Recorder keeps a record of the group's official answers in his or her workbook. This allows the instructor to keep tabs on all the groups and encourages groups to come to consensus about each answer.

Reflector   Observes and comments on group dynamics and behavior with respect to the learning process. These observations should be made to the manager on a regular basis (no more than 15 minutes between reports) in an effort to constantly improve group performance. The reflector may be called upon to report to the group (or the entire class) about how well the group is operating (or what needs improvement) and why.

Other roles can be used, but are not described here (e.g., Model Builder, Encourager, Checker).

Classroom Management

Group Folder (Collecting and Distributing Materials)
In most POGIL classes each group has a folder. The group folder is used to return graded quizzes or other work at the start of class. Folders are placed on the head table, and a group representative comes and gets the folder before class begins. At the end of class, group managers may be asked to place work to be graded in the group folder before it is returned.

Instructor's Role as Facilitator
The guiding principle in facilitation is that **students gain far more from correcting their own answers** than from being corrected by the instructor. Do not interpret this as meaning that instructors should refuse to answer questions. A successful facilitator will give students the minimum amount of information required for them to correct their own mistake, while being supportive and encouraging. If the instructor gives too much help (e.g. frequently answers the question "Is this correct?") students will quickly get the message that they cannot trust their own self-assessment skills, and must always get expert confirmation of their conclusions. This is poor training for success on quizzes, exams, and especially for the real world, where there is no answer key.

---

[12] Farrell, J.J.; R.S. Moog, and J.N. Spencer. "A Guided Inquiry General Chemistry Course." *J. Chem. Ed.* Vol. 76, No. 4, April 1999. p 570-574.

If it seems likely a subsequent question will cause a group to correct their own error, the best course of action may be to *not* intervene. In such cases, the facilitator should return to the group to make sure the problem was solved without intervention.

The following table gives examples of some common interventions that can help students self-correct, without developing dependency on instructor confirmation.

Overview of Instructor Facilitation
Observe <u>each</u> group to monitor progress and keep track of which groups are having problems. Focus on the key questions (marked with a key in the margin).
Intervene with <u>individual</u> groups when necessary. For problems affecting more than two groups, a whole-class intervention may be preferable.
Examples of Group Level Interventions: Ask a group to… • Have the manager read the question out loud • Assess confidence in their answer • Check if all group answers match • Read the next question or model • Consult with a neighboring group • Practice using the bullets above *without prompting by the instructor* when stuck or looking for confirmation of an answer
Examples of Whole-class Interventions: • Ask group(s) to write their answer to a key question on the board (or give a group an overhead and marker) • Call on a group to critique or explain an answer • Call on a group to report answer to the whole class • Vote on competing answers • Instructor mini-lecture

Another key principle of facilitation is that **students critically evaluate comments made by other students**; however, many students try to memorize (without reflection) what the *teacher* says. Whenever possible, use student-offered explanations to provide examples or stimulate discussion. An answer offered by a student may be identical to the correct answer offered by the instructor, yet the former will win more scrutiny from students.

A common pitfall of facilitation is for the instructor to fall into a dialogue with just one member of a group. This most often happens when the instructor is responding to a question posed by one member of a group. When the instructor approaches a group with a question, the first thing to do is make sure all members of the group have already considered this question. One good technique is for the instructor to ask a group member (other than the one with his hand up) to state the question on behalf of the group.

It is also tempting for the instructor to engage in a private dialogue with a student who is behind and needs extra help. Involving the rest of the group is worth the time and energy it takes, even if it means slowing down other group members. As the course progresses, the power of learning by teaching will become apparent, and most students will figure out that a less confident student who asks lots of questions is an asset to the group. The instructor may need to jump start group interdependence by, for example, asking a stronger member to explain a concept to a group mate.

Instructor interactions with the group send powerful messages about how the group should interact. **The instructor should model how students should interact, asking for explanations and reasoning.** When two groups have the same problem, the instructor can ask a representative of each group to switch chairs. Or, if they are close enough, the instructor can point to the other group and say: "Those guys are stuck on the same question that you are discussing. I think you will find their answer very interesting." In the future, students may consult a nearby group without being prompted by the instructor.

Summary Lectures and Other Teacher Talk

Every POGIL class typically includes several short periods of teacher talk (lecture). Class often starts with a brief (three-minute) overview of the previous class, and a one-minute preview of the upcoming class. A general rule is that the larger the class, the more time you will spend at the front of the room, speaking to the class as a whole. If several groups are having the same problem, it can be more efficient to stop group work and address the problem as a class.

Whole-class (plenary) teacher talk can also be useful for speeding student work on an activity. If progress is too slow, interrupting group work and giving away an answer shuts down further discussion since the "expert" has ruled in with the "correct" answer. Be aware that instructor input may have this effect, and that each attempt to speed the class along has a cost. The instructor must decide which is more beneficial at that moment: to move students toward the main concept waiting at the end of an activity, or to allow for discussion that will deepen student understanding of this concept.

If the instructor intervenes too frequently this will reinforce the commonly held student belief that no answer is correct until confirmed by the instructor. With that said, however, revealing too little also has a cost. Especially at the start of a course, student frustration thresholds may be low. The number one complaint from POGIL students continues to be "the instructor refuses to answer my questions." Explaining the pedagogy (essentially telling students "it is for your own good when I force you to figure out answers for yourselves") may only intensify anxiety and frustration. The instructor should acknowledge the tricky spot, and perhaps give a hint. This lets the students know he wants to help. Ideally, the instructor gets to know each student's frustration threshold so he can give only the minimum information required for a student to figure out the concept for herself.

According to Piaget, frustration is an important part of learning.[7] The trick is to push students, but not make them snap. Some students have high tolerance for the cognitive discomfort inherent in learning; others do not. It is deeply gratifying for an instructor to know students at this level, and very exciting for everyone involved when this knowledge helps students go beyond their perceived limitations.

Student Self-Management of Class Time

The activities in this workbook are designed to be completed in one class period. For fifty-minute classes, more work will have to be done outside class than for longer classes. The instructor should not let poor progress on an activity cause work on that activity to roll over into the next class period.

The instructor must make it clear that students are <u>responsible for finishing each activity before the start of the next class</u>. Challenging them to complete the activity during or after class will do two things: 1) Encourage students to manage class time effectively and 2) encourage out-of-class group work.

If an individual student or a group is consistently behind the rest of the class, a good strategy is for such students to preview the activity by reading the Model and Questions before class.

Daily Quiz

It is strongly encouraged that each POGIL class start with a one-question (three-minute) quiz. Ideally, the quiz should cover one important concept that was developed during the *previous* class meeting(s), but serves as foundational knowledge for the *upcoming* activity. At the end of the quiz, the instructor should (very briefly) go over the quiz. Otherwise, students will spend the first five minutes of group work time discussing the answer to the quiz.

Group Take-Home Exams

A great way to foster group interdependence is to assign a group take-home exam. For such an exercise, it can be better to let students form their own groups of either three or four. By requiring that each group turn in only one copy of the exam, the instructor forces students to come to consensus. A good rule is to specify that groups can consult any published work, but they can only talk to their group-mates or the instructor about the exam. After such an exercise there is an increase in out-of-class group work. For this reason, it is good to assign a group take-home early in the course.

Use of a Traditional Textbook

These POGIL materials are intended to be used in conjunction with a traditional textbook. Since most textbooks are written at a fairly high level, reading in the text is not usually a good first introduction to a topic; however, students who have already been introduced to a concept via an activity can gain enormously from reading about it in a textbook. For this reason, it is best to assign reading on a topic *after* the students have completed the relevant ChemActivity.

Use of the text as a reference during class is not recommended, especially at the start of the course. As the course progresses, students who have spent significant time using the text outside of class may find it useful as a reference during class.

Most successful students read in the text after each class to try and confirm their answers to the in-class Construct Your Understanding Questions. If you have ever waited for tomorrow's paper to check your crossword answers, you know that "expert" answers are extremely interesting only if you have spent time struggling to come up with your own, possibly incomplete answers.

Assessment: Improving Instructor and Student Performance

Assessment of Course and Instructor by Students

It is strongly recommended that the instructor ask students for feedback about the course sometime in the first half of the course. (So there is time to make adjustments.) One strategy is to give students ten minutes of class time to describe:

- at least one strength of the course, and how this strength helps them learn
- at least one area for improvement, and (if possible) suggest possible changes
- any other insights about teaching and learning.

Another way to collect this information is to have students email responses to a secretary, who can print the emails, cut off the headers and give the instructor the now-anonymous feedback. To ensure full participation, the secretary should keep a list so responders can be awarded bonus points. If no bonus points are offered for an out-of-class feedback exercise, compliance is likely to be low and skewed toward students with extreme viewpoints (positive and negative).

Assessment of Group by Students

Several times, particularly at the start of a semester, it is useful for each group do a self-assessment. This is the same as above, except that students write down one strength of their *group*, and one area for improvement in their *group*. Each group member should take one minute to share their assessment with the rest of the group. Written reports should be turned into the instructor. If time allows, the instructor can ask for volunteers to report key insights to the whole class.

Self-Assessment of Group Participation by Students

Another type of self-assessment is to distribute the table below with the following instructions:

For each __row__ on the table, circle the statement that best describes __YOU__ in terms of participation in your group. (Not to be collected or graded.)

Excellent (4)	Good (3)	Fair (2)	Poor (1)
Lead *and* share the lead without dominating	Lead but dominate a bit	Follow but never lead	Actively resist group goals
Actively pace group so everyone is on the same question & finish on time	Aware of time issues but don't actively work to keep group together & on pace	Don' t think much about group progress or timing	waste lots of group time, fall behind, or work ahead
Stay on task and keep others on task	Keep self on task	Sometimes get group off task	Often get group off task
Actively create environment where everyone feels comfortable participating	Try to engage others in a helpful and friendly way	Rarely initiate interactions, but respond in a friendly way when others initiate	Observe silently, and offer little when others try to engage you
Express disagreement directly and constructively	Usually express disagreement directly	Avoid confrontation even when angry or frustrated	Let negative emotions get in the way of team goals
Enthusiastic and positive	Moderately enthusiastic	Show little enthusiasm	Negative or unenthusiastic
Always come prepared	Usually prepared	Occasionally unprepared	Usually unprepared

Encouraging Students to Answer the Central Meta-Question

On the surface, the students' job is to construct and refine answers to the Construct Your Understanding Questions; however, underlying most every question is another question:

> *"What is the purpose of this question; what point, distinction or concept is the author trying to convey by asking us this question; what am I supposed to learn about the Model from this question?"*

Answering this *meta*-question requires a student to see the forest from within the trees, and to get inside the author's mind. The instructor may occasionally ask students to explicitly answer the meta-question, but students who get in the habit of doing this will gain a deeper understanding of the material and achieve greater success in the course.

Final Words of Advice

Most students and instructors need at least a few class periods to find their rhythm with POGIL, but by mid-semester it should start to come together. If it does not feel right, seek help early. Instructors can contact the POGIL Project Office (www.pogil.org), and be matched with an experienced POGIL mentor. Students should address their concerns to their instructor during office hours (not during class).

For instructors, the hardest part appears to be figuring out when and how to interrupt. At the beginning err on the side of interrupting more often and saying more. It is easier to step students back from overdependence than to repair their frustration with the instructor.

For students, the hardest part is getting over the anxiety surrounding the question "How do I know if my answers are the right answers." Get to know the feeling of knowing you are right. Keep asking questions, reading and doing problems until you are confident you understand. Now you are ready for the exam.

Frequently Asked Questions *(with answers)*

FAQ For the Instructor (Professor, Peer Leader, or TA)

How do I deal with room and seating issues such as fixed seating?

An ideal POGIL environment has 20 students in five groups of four, each sitting at a square table with four chairs. Few people have the luxury of such a classroom. Most everyone has to make some kind of accommodation to their specific environment.

Two common arrangements are side-arm desks, and rows of tables with chairs facing the front. With side-arm desks, the main problem is that some groups are reluctant to make their circle tight enough for effective group work. Students may require encouragement to scoot closer together. With rows of tables all on the same level, turn half the chairs around to face the back.

Fixed seating in a lecture hall presents special challenges.

- In a fixed seat lecture hall groups of three work best. Groups of four work also, but often function as two loosely affiliated groups of two.

- If your lecture hall is large enough, leave every third row open so the instructor can walk down the empty rows and field questions from above and below without squeezing past student's knees and tripping on their backpacks.

- Students should work in a way that is comfortable to them (without violating fire codes). In a lecture hall some students choose to kneel on their seats to interact with group mates in the row behind. Others groups have preferred to sit on the floor at the front or back.

How are TAs and peer leaders best trained in the use of POGIL?

Many former POGIL students make excellent facilitators, even without additional training. If this is a possibility, it is always a good strategy to select peer leaders from among last year's POGIL students in the same course.

When this is not possible, an experienced teacher (e.g. the lead instructor in the course) with an understanding of POGIL should meet with the TAs prior to each week for the first few weeks of the course and facilitate a "master class" in which TAs do the activity for that week. Put TAs in groups of four and have them discuss each question. Though the content may be easy for many TAs, it is still worth taking the time to go through the whole activity since it allows TAs to address any questions, and the master teacher to model POGIL facilitation techniques.

After a few weeks of modeling POGIL facilitation during TA-meetings, it may make sense to have the TAs take turns running the TA meeting. Having TAs go through the activity on their own is not nearly as beneficial. If a quiz will be given in the recitation section, the TA meeting should begin with the TAs taking a sample quiz.

How do I recruit undergraduate peer instructors for my Peer Led Team Learning sessions?

Recruiting undergrad peer leaders to facilitate *next year's* peer led sessions is as easy as simply asking for volunteers from among your current "A" students. Students who enjoyed being POGIL students are often eager to continue as peer leaders. Other motivating factors include the opportunity to review for the MCAT, PCAT, etc., and the opportunity to receive teacher training and mentoring from the professor. In situations where budgetary constraints preclude payment of peer leaders, these factors can be sufficient motivators with or without course credit. Of course, recruiting from your past POGIL students is impossible the first time you teach a POGIL course at a given institution.

What should I (the instructor) do the first day in class?

A good strategy is to give a three-minute mini-lecture on key expectations (e.g. work together, complete the activity by next class, be prepared for a quiz), then launch immediately into the first activity. Organic chemistry students are generally anxious to focus on the material that is going to be on the exam. Though understanding POGIL will help them do this, the best way for students to learn about POGIL is by doing the activity. Additionally, it is somewhat hypocritical to lecture students on a lecture-less teaching method.

Is it better to let students select their own groups, or for the instructor to assign groups?

There is some evidence that homogeneous groups lead to the most rapid adjustment to group work. On the other side of this ledger is the fact that "learning to work with a diverse range of people" is frequently cited as a key skill that students need for success in today's world. It is up to the instructor to balance these competing goals.

Why do my students seem to have so many misconceptions?

POGIL instructors get to see what their students know and don't know. After lecturing for years, and assuming that certain concepts are straightforward, instructors are surprised at what their students think, especially about topics they always assumed were covered adequately in prerequisite classes. POGIL instructors get to watch students struggle with concepts and grow their understanding through some very sensitive (and even ugly) stages – students' initial conclusions and the misconceptions carried from prior classes can be fascinating and unexpected. Many of these common misconceptions are compiled at the end of each activity in the **Common Points of Confusion** section, and can be useful to both students and instructors.

Does using POGIL take away the fun of being the Sage on the Stage?

No. Though POGIL instructors do not talk as much as lecturers, the talking they do can be very satisfying. When the instructor is not talking to the whole class, she is walking around the room telling stories, giving analogies, and participating in discussions with student groups. I became a professor because I like to explain, profess, and otherwise share my love and understanding of organic chemistry. You do not give that up by using POGIL. In fact, you may be surprised to find your five-minute mini-lecture sets heads nodding (in agreement, not to sleep), and leaves eyes blazing (instead of glazing).

How do I (the instructor) deal with fast vs. slow groups?

There are a number of facilitation techniques to deal with the diversity of group work rates.

- Slow groups are often slowed by one or two students who like to work slowly and think hard about every question. One solution is for these more careful students to preview the activity before class. After trying this once, students usually find it self-evident that this strategy allows them to get so much more out of group discussion and teacher talk.

- Groups that go really fast often miss the point of some key questions. It can help to simply point this out.
- A group of very bright and confident students may work quickly through an activity and get everything they are supposed to get from it.
    - It can be fun to give this group an auxiliary question of a more challenging nature.
    - Alternatively, the instructor can ask such a group to stand up and walk around as co-facilitators, checking the conclusions of other groups and learning by teaching.

What should I (the instructor) do when I see a wrong answer?

Students learn more when they find their own error and correct it with the minimum instructor intervention. The following are techniques you can use to coax students toward understanding.

- Simply having one student in the group read the question out loud can help the group see an error of interpretation.
- Point students to the relevant section of the Model.
- Often one student in a group will have a correct answer. Rather than pointing this out, the instructor can simply point out that there is a difference of opinion within the group.
- If a whole group is wrong, a great technique is to have one member of the wrong group switch chairs (temporarily) with a member from a correct group, not telling them which group is wrong and which is right so both groups critically evaluate both answers.
- If a large number of groups are off target on a question the instructor should lead a whole class discussion. One good format is to have a representative from several (or all) groups put their answer on the board. (If there is limited board space, pass out blank overhead sheets and overhead pens.) Voting usually reveals the strongest answer, and can lead into a useful discussion of why the wrong answers are wrong.

How should the instructor answer the question "Is our answer right?"

Don't be abrupt and say "I can't answer that." Instead, ask students to explain their answer, or why they think they are right or wrong. In every situation, encourage students to talk about what they know and don't know. This helps the instructor hone in on the misconception or bad assumption.

If class is about to end, or the group seems really frustrated, the instructor might consider simply giving the answer, even though this is almost always less beneficial than helping the students come to their own valid conclusion. If a group or individual is so frustrated that they disengage, this can have long term implications that far outweigh damage to the learning process caused by giving away an answer now and then.

How do I (the instructor) encourage internal processing of questions?

When approaching a student with her hand up, a good strategy is to point to another student in the group and say "Do you know what her question is?" At first, the answer might always be no, but students will quickly learn that most questions can be answered within the group without instructor involvement. If the question has not been discussed yet, you can offer to come back (or if there is time) stay and observe while the original questioner explains her question to the group.

What should I (the instructor) do when I make a mistake?

First, and most importantly, try not to feel bad. It happens to even the most experience instructors. Thinking on your feet in a POGIL classroom is in many ways more challenging than lecturing. Because students are thinking deeply about the material, they will invariably ask questions that you have never considered, and make reasonable arguments leading to conclusions you know are contrary to experiment. Occasionally, instructors get confused by one of these and end up confirming an incorrect conclusion or making an incorrect statement that leads students astray.

The climate of the classroom, and the scope of the error should determine how the instructor unwinds an error, but keep any correction simple. There is a strong temptation to carefully detail (in your own defense) the logic of even the most esoteric error. The result is to draw attention to the issue and give students the impression that this issue is more important than it actually is.

What should I (the instructor) do in the last 5-10 minutes of class?

This is a good time for group self-evaluation, especially at the start of the course or on a day when students gained a strong grasp of the material. Such discussion helps groups work through issues that might hinder them going forward.

On other days, it is best to facilitate a whole-class content summation. This can be presented wholly by the instructor or (even better) by students. The more challenging the material, the more important it is to have a content summary at the end.

Are there POGIL activities other than the ones in this book?

Organic Chemistry: A Guided Inquiry *for Recitation* consists of two volumes. Volume 1 covers key Organic 1 topics, Volume 2 covers key Organic 2 topics. A full-course set of materials (called simply, *Organic Chemistry: A Guided Inquiry*) is also available and includes many topics not covered in either of these slim recitation volumes is. The full-course set contains activities on most first year organic topics, enabling an instructor to use POGIL in nearly every class meetings (see next topic).

POGIL activities are available for most chemistry courses (including general, physical, analytical, and biochemistry), and for some courses in biology, mathematics, and engineering. More information is available at www.pogil.org.

Can one do POGIL during every class (instead of just once a week)?

Many POGIL courses consist of group work every day, instead of just in recitation. Such an approach has been employed successfully in both small and large (>350 students) classrooms. In large classrooms (>50 students) POGIL is usually used in conjunction with electronic classroom response devices ("clickers"). More information on the use of POGIL in large classes can be found at www.pogil.org/straumanis.

FAQ for Students and Instructors

Wouldn't it be easier for students to learn the "right" answers if the instructor posted them?

Of course it would be easier if the goal was simply to have students memorize answers to questions. However, the Construct Your Understanding Questions are far too easy to put on an exam. (They are designed to be your introduction to a topic). It turns out that memorizing answers to simple questions does not help you answer harder, more conceptual questions. HOWEVER, wrestling with simple questions and coming up with your own answers is a fantastic entry into a topic, and the best preparation for your further study via homework, lectures, and quiz and exam questions.

Another way of saying this is: "When a student reads a question and doesn't immediately know the answer, the tendency is to immediately look at the answer key if it's available. This hurts the learning process since most learning takes place as you try to figure out the answer to a question by talking to others, reading the textbook, revisiting the activity, etc. Wrestling with the questions is an important part of the learning process in this class, and giving out the answers severely short circuits this process.

If I (a student) understood the activity, do I still need to do the homework and reading?

Some students leave their POGIL class feeling they have learned enough that they do not need to do the homework and reading. No matter how much you feel you understood the activity during class—class is just the start. The homework and reading often contain important extensions of the basic concepts covered in class. The instructor can quickly convince students that skipping homework was an error by pulling the quiz from something that only appeared in the reading or homework. Because POGIL activities are designed to be a student's first introduction to a course, successfully answering the in-class Construct Your Understanding Questions does not, alone, constitute mastery of a topic.

What is a Recorder's Notebook, and how is it used?

Especially if there was no time for a content summary at the end of class, it can be useful to require students or groups to keep a **Recorders Notebook**. In this, students write down the two or three most important concepts from the day's activity. The instructor can spot check these (credit/no credit) during the quiz in the next class.

What if students don't finish the activity in class?

The instructor must make it clear to students that they must finish the whole activity (including homework and reading) before the next class, either on their own, or as a group. This knowledge will help students manage their time, including making a start on the activity before class if necessary.

How does POGIL affect students with learning differences: dyslexia, ADHD, ADD, etc.?

Because POGIL is student centered, it allows students to adapt the classroom to their strengths. Students who are verbal learners can talk it out. Students who learn by listening can listen. Students who learn by reading can read...etc. This same flexibility helps hearing and visually impaired students adapt to a POGIL environment.

Why spend energy improving learning and process skills? (Will they be on the exam?)

A POGIL class is as much about critical thinking, problem solving and group work as it is about content (e.g. organic chemistry). Of course, the two go hand-in-hand. The better a student's learning skills, the more she will learn. It is also the case that in today's world such skills are highly prized. You cannot be a successful scientist or professional without these skills. Alumni of POGIL classes report that the process skills they learned helped them succeed in graduate school, jobs, medical school etc. As an aside, a common comment is that POGIL is a great topic to raise in an interview. POGIL is new enough that many interviewers have never heard of it, and it fits well with the big push at many institutions for creative problem solvers who work well as part of a team.

Will students like POGIL?

Student grumbling about POGIL in the first week of the course is normal, especially the first time POGIL is used for a given course or at a given institution. Any new classroom method is met with resistance by some students. Most students quickly adjust, and by the end of the course only about 10% are negative about POGIL.

Interestingly, at the start of a course the strongest students are the most critical, but this quickly turns around. By the end, the strongest students become the biggest advocates of POGIL. For example, in a recent anonymous survey of 240 POGIL students none (0%) of the 25 students who received an A in the course were negative about POGIL, and only 2 (6%) of the 31 B+/A- students were negative. The majority (80%) of the students who gave POGIL a negative review received a C or lower in the course.

How did the author get involved in POGIL?

When I was a student at Oberlin College I had excellent teachers, but the format was straight lecture. It was my experiences *after* each class that shaped my contributions to what eventually became POGIL.

Every day in my organic chemistry class I wrote down everything the professor put on the board so I could figure it out later. I was lucky to have two other students in the class who were on the soccer team. On the bus going to away games we would sit in the back and work together. Home game weeks we would meet in a study room in the library. (We tried meeting in a dorm room once, but it had a television and we ended up watching Magnum PI.)

In the group sessions we went through each example in our notes and tried to answer the question: "What concept or distinction is the professor trying to convey with this example?" Then we did the homework problems together. If we came across a question we couldn't answer ourselves we asked another group of students who studied in an adjacent room in the library (or, when on the bus, we asked a teammate who had taken the course the year before). If this didn't help, we wrote down our question to ask the professor in the morning.

It struck me that many of the other successful students in the class were doing the same thing as we were, but that there were also many students who were struggling, working on their own.

Later, in graduate school at Stanford University, I encountered the work of P. Uri Treisman and others, which explained why my study group in college was so effective. I had my own recitation sections and I began to experiment with ways to encourage study group behavior in my students during class. A key element of this was the use of guided inquiry: using leading questions to guide students toward understanding of a concept.

About this same time I met Professor Rick Moog, of Franklin & Marshall College, while he was visiting his former graduate research advisor (also at Stanford). He invited me to attend his talk the next day at the American Chemical Society Meeting in San Francisco. It turned out that he had been thinking about similar issues: group learning and guided inquiry. I found Rick was way ahead of me and was about to publish guided inquiry activities for general chemistry. I was so stimulated by his fifteen minute talk that as soon as he finished, my hand shot up and I asked: "Who is writing similar activities for organic chemistry?" He looked me in the eye and said "You are."

Inspired by this, I began work immediately on what eventually became Organic Chemistry: A Guided Inquiry. A few years later, a generous grant from the National Science Foundation allowed us (along with five others) to form the POGIL Project, which has been the focus of our now fourteen-year old collaboration.

My teaching has changed and evolved during the past 20 years, but my primary goal remains the same: to create a classroom environment where students work together in pursuit of a deeper understanding of organic chemistry, and in the process learn how to exceed their own perceived limitations.

Andrei Straumanis, October 2010

Index of Frequently Asked Questions

(answers in the previous section)

Questions for the Instructor (Professor, Peer Leader, or TA)

How do I deal with room and seating issues such as fixed seating?

How are TAs and peer leaders best trained in the use of POGIL?

How do I recruit undergraduate peer instructors for my Peer Led Team Learning sessions?

What should I (the instructor) do the first day in class?

Is it better to let students select their own groups, or for the instructor to assign groups?

Why do my students seem to have so many misconceptions?

Does using POGIL take away the fun of being the Sage on the Stage?

How do I (the instructor) deal with fast vs. slow groups?

What should I (the instructor) do when I see a wrong answer?

How should the instructor answer the question "Is our answer right?"

How do I (the instructor) encourage internal processing of questions?

What should I (the instructor) do when I make a mistake?

What should I (the instructor) do in the last 5-10 minutes of class?

Are there POGIL activities other than the ones in this book?

Can one do POGIL during every class (instead of just once a week)?

Questions for Students and Instructors

Wouldn't it be easier for students to learn the "right" answers if the instructor posted them?

If I (a student) understood the activity, do I still need to do the homework and reading?

What is a Recorder's Notebook, and how is it used?

What if students don't finish the activity in class?

How does POGIL affect students with learning differences: dyslexia, ADHD, ADD, etc.?

Why spend energy improving learning and process skills? (Will they be on the exam?)

Will students like POGIL?

How did the author get involved in POGIL?

Notes

ChemActivity 1: Aromaticity

(What structures are likely to exhibit the special stability known as **aromaticity**?)

Model 1: Cyclic Conjugated π System ("Racetrack" Analogy)

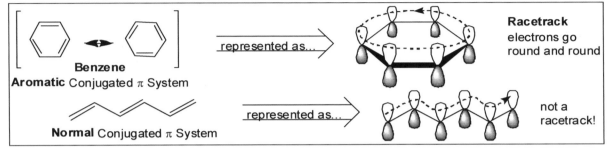

- Certain cyclic molecules exhibit an almost magical stability, attributable to lower than expected potential energy. These molecules are called **aromatic molecules**.

- Before it was understood, this "magical stability" was named **aromaticity** because many molecules in this class have a strong **aroma** (smell).

- The "racetrack" analogy explains this stability as follows: electrons can race endlessly around the "racetrack" of conjugated *p* orbitals. This allows them to adopt a longer wavelength, and thus a lower energy. We call such a "racetrack" of *p* orbitals a **cyclic conjugated π system**.

<u>**Not all cyclic conjugated π systems are aromatic**</u>. Of the structures below, only those that are labeled "**Aromatic**" display the special stability associated with aromaticity.

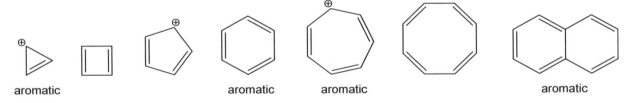

Figure 1.1: Examples of cyclic conjugated π systems

Construct Your Understanding Questions (to do in class)

1. Confirm that there is a *p* orbital (empty or π-bonded) at each carbon in Figure 1.1.

   a. Find the largest possible cyclic conjugated π system ("racetrack") in each molecule in Figure 1.1, and label it with the total number of electrons involved in π bonds of that "racetrack." (For these molecules, this is also the total number of **π electrons** in that racetrack.)

   b. Certain "magic numbers" of π electrons give rise to aromaticity. According to the data in Figure 1.1, circle the "magic numbers" in the list below.

   2   4   6   8   10   12   14   16   18   20

   c. Continue the series past 10 electrons: Which of 12, 14, 16, 18, and 20 appear to fit the pattern and are expected to be "magic" numbers that allow aromaticity? Circle these above.

d. For a molecule to be aromatic, it need only *contain* a cyclic conjugated π system (racetrack) with a "magic" number of electrons. (That is, there can be parts of an aromatic molecule that are not involved in the "racetrack.") <u>Four</u> of the following molecules are aromatic. Identify these and explain your reasoning.

Memorization Task 1.1: Hückel's Rule

In 1938 German chemist Erich Hückel noticed that molecules with 2, 6, 10, 14, 18, 22, etc., electrons in a cyclic conjugated π system displayed the special properties of aromatic molecules.

This came to be called "**Hückel's Rule**," which states: "**a cyclic, continuous, conjugated π system exhibits aromaticity (special stability) if it contains 4n + 2 π electrons** (where n = 0, 1, 2, 3…etc.)."

---

2. (Check your work.) Are your answers above page consistent with Memorization Task 1.1?

3. Consider the orbital representations of benzene below.

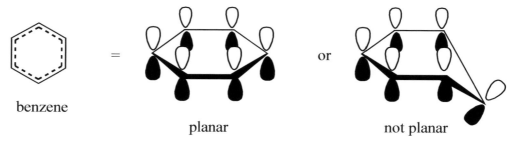

a. Which π system (planar or not planar) has the better *p* orbital racetrack?

b. Is your conclusion consistent with the fact that rings capable of aromaticity only exhibit special stability when they are planar? (Benzene is therefore always planar.)

4. Identify any sp^3 hybridized atom in the molecules below, and construct an explanation for why an sp^3 hybridized atom in a ring interrupts the "p orbital racetrack" along that path in the ring. (None of the examples below are aromatic.)

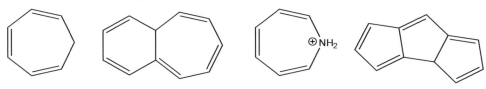

Memorization Task 1.2: Rule for determining when a lone pair will reside in a _p_ orbital

A lone pair on an atom in a cyclic conjugated π system will occupy a _p_ orbital (and become a part of the π system of the molecule) **if doing so makes the molecule aromatic**.

---

5. Predict which <u>one</u> of the two molecules at right will be aromatic and explain your reasoning.

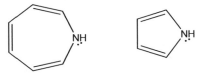

6. Construct an explanation for why the lone pair on N in the molecule below CANNOT participate in the π system of the molecule (even though the molecule, as drawn, has four π electrons, and so adding two more would give it six). _Hint_: p orbitals in an aromatic ring _must_ be parallel.

7. In the following molecule, one lone pair resides in one type of orbital and the other lone pair resides in a _different_ kind of orbital.

    a. State what type of orbital (s, p, sp, sp^2 or sp^3) each lone pair resides in and explain your reasoning.

    b. (Check your work) Is your reasoning to the previous question consistent with the fact that placing both lone pairs in p orbitals would generate large amounts of electron-electron repulsion since the p orbital _not_ participating in the aromatic system would point directly into the ring, with less than 90° separating it from the other bonds.

8.   Indicate the type of orbital holding each lone pair, and label each molecule that is aromatic.

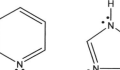

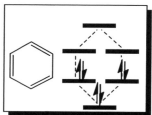 9.   Review: For a molecule to be aromatic it must have a π system that …(***fill in the blanks***).

      (1)   is cyclic

      (2)   has a continuous set of _____ *p* orbitals (no sp^3 carbons in the ring)

      (3)   has _____ π electrons (2 or 6 or 10 or 14, etc.)

      (4)   and has an overall _____ geometry allowing the *p* orbitals to be parallel.

Model 2: Molecular Orbital Diagrams of Cyclic π Systems

The π molecular orbitral diagram (**π MO diagram**) of a cyclic conjugated π system takes the shape of the ring involved, with one vertex pointing down. For example…

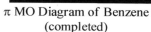 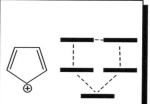  

π MO Diagram of Benzene
(completed)

π MO Diagrams of Other Cyclic Conjugated π Systems
(electrons not filled in yet)

- Molecular orbitals on a MO diagram with the same energy are called **degenerate orbitals**.

- All degenerate orbitals must be half-filled before any one is filled.

Memorization Task 1.3: Radical Character

A species with an unpaired electron is said to have "**radical character.**" Usually a species with radical character is very reactive.

Extend Your Understanding Questions (to do in or out of class)

10. Circle a pair of **degenerate** molecular orbitals (there are two) on the MO diagram of benzene.

11. Based on the structure shown, add the appropriate number of π electrons to the three other π MO diagrams in Model 2.

12. Label two compounds in Model 2 with the words "**has radical character-very reactive.**"

Memorization Task 1.4: Anti-Aromatic Molecules

A molecule is **anti-aromatic** (and highly reactive) if it exhibits all characteristics of aromaticity EXCEPT that it has 4n (instead of 4n+2) π electrons (e.g., 4, 8, 12, etc.).

---

13. Cyclobutadiene can be isolated and studied at temperatures lower than –200°C. At temperatures above –200°C it immediately reacts with itself to form other more stable products. Use molecular orbital theory to explain why **cyclobutadiene** is so unstable.

    Cyclobutadiene (very unstable)

14. List the key attributes of cyclobutadiene that make it anti-aromatic.

15. (Check your work.) Compare your answer above to your answer to Question 9.

16. Cyclooctatetraene is quite stable at room temperature. However, close examination of the structure reveals that **it is not planar**. This can be explained by optimization of bond angles.

Cyclooctatetraene

Cyclooctatetraene in its
Preferred Conformation

    a. Use MO theory to construct an argument **not based on bond angle optimization** that explains why cyclooctatetraene is not planar.

    b. Cyclooctatetraene is considered **non-aromatic.** This means it is neither aromatic nor anti-aromatic. Explain.

17. Cyclobutadiene is one of the few molecules that is unable to "avoid" being anti-aromatic. Explain.

Memorization Task 1.5: Most molecules are non-aromatic & very few are anti-aromatic.

Cyclobutadiene is too small to adopt a nonplanar geometry. Most other cyclic conjugated π systems with 4n π electrons can flex to adopt a conformation that is not planar, or adopt a hybridization state that breaks the conjugated π system, thereby avoiding the instability that comes with anti-aromaticity.

Bottom line: There are <u>very few anti-aromatic molecules</u> and many aromatic molecules, but the majority of molecules are neither aromatic nor anti-aromatic (these are called **<u>non-aromatic</u>**).

Confirm Your Understanding Questions (to do at home)

18. Construct an explanation for why the N on the following molecule is expected to be *sp^3* hybridized.

19. Draw an example of each of the following that does not already appear in this activity: an anti-aromatic molecule, an aromatic molecule, and a non-aromatic molecule.

20. Only one of the following molecules is aromatic. Label it **aromatic**.

21. Indicate the largest aromatic ring in each of the following aromatic molecules. The first two are done for you. Note that the second example has two **aromatic rings**. Even though these two aromatic rings are conjugated together, they are NOT considered a single aromatic system (The molecule has two separate aromatic "racetracks.").

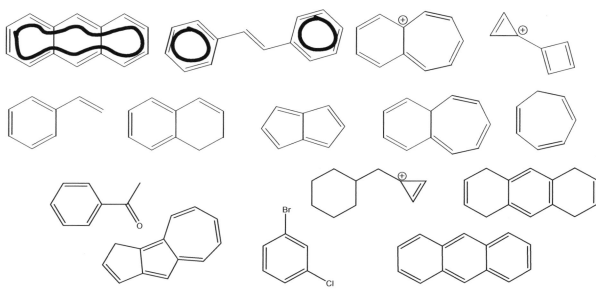

22. Draw the enol form of the following ketone, and construct an explanation for why the enol form is favored in this special case. (Recall that normally the keto form is lower in potential energy and therefore favored.)

keto form   tautomerize   enol form
(favored in this case!)

23. For a set of resonance structures, the members of the set that are aromatic are considered lower-potential-energy arrangements of the electrons. These aromatic resonance structures are therefore considered "more important" than their non-aromatic counterparts. For each of the following, draw all important resonance structures, and mark the ones that are more important (contain an aromatic ring).

24. Explain why each of the following structures (as shown) is NOT aromatic.

25. The following ketone has a large amount of + charge on carbon and – charge on oxygen. This is because the so-called zwitterionic form of the ketone is much more favorable than you might expect given that it is charged. (A zwitterion is a compound with both a + and a – charge.) Construct an explanation for why the compound has a large amount of zwitterionic character.

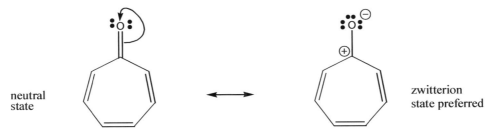

neutral
state

zwitterion
state preferred

26. The molecule on the left has a lone pair that resides in a *p* orbital despite the fact that in a *p* orbital the lone pair resides closer (90° vs. 109.5°) to N-H and N-C sigma-bonding electrons. Construct an explanation for why this lone pair resides in a *p* orbital.

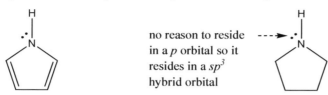

no reason to reside in a *p* orbital so it resides in a *sp³* hybrid orbital

27. Indicate the type of orbital holding each lone pair, and label each molecule that is aromatic.

pyridine   imidazole   furan   pyrrole   cyclopentadienyl anion (abbreviated "cp")

pyrimidine   quinoline   purine   pyrrole anion

28. All of the "heterocycles" in the previous question are commonly encountered in nature or the laboratory. (A heterocycle is any cyclic molecule containing at least one non-carbon atom. If you continue your study of organic chemistry beyond this course you will likely become familiar with these common heterocycles.) <u>Construct an explanation for why each of the heterocycles shown above is common.</u>

29. Draw an energy diagram showing both of the following reactions on the same reaction coordinate, and construct an explanation for why the pK_a of cyclopentadiene is much lower than the pK_a of cyclopentane.

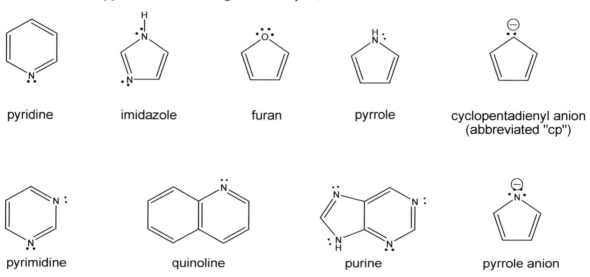

uphill by +50pK_a units

$pK_a = 50$

uphill by +16 pK_a units

$pK_a = 16$

30. Explain why carbocation B is formed preferentially when water leaves. Then use a curved arrow to show the mechanism by which B is formed from A.

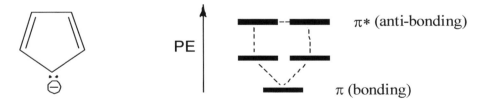

A   B

31. A benzyl carbocation can rearrange to the ion shown below right (called a tropylium ion).

   a. Use a curved arrows (bouncing curved arrows are useful here) to show the mechanism of this two-step carbocation rearrangement.

   b. What is the driving force for each step in this rearrangement?

32. When an organometallic chemist needs a large anion that will make a strong bond to a metal, a common choice is the anion of pentamethylcyclopentadiene (pentamethylcyclopentadienyl anion).

   a. Draw pentamethylcyclopentadienyl anion, and explain why it is a stable anion.

   b. Construct an explanation for why pentamethylcyclopentadienyl anion was nicknamed "**cp*.**"

33. The anion below left, has two π bonds that contain four π electrons. Ignore the lone pair on the anionic carbon for now, and place the four electrons from the two π bonds in the MO diagram.

PE   π* (anti-bonding)

π (bonding)

   a. According to your MO diagram with four electrons, does this anion have radical character?

   b. NOW assume the lone pair on this anion resides in a *p* orbital, and **add two more electrons to your MO diagram** to reflect that the lone pair electrons are part of the π system.

   c. Do you expect the lone pair on this anion to reside in a *p* orbital as we assumed in part b? Explain your reasoning.

   d. Is this cyclic anion expected to be aromatic?

34. Draw an MO diagram for tropylium ion (structure shown on the previous page), and fill in the π electrons showing the ground state of this ion.

35. The first half of Hückels' Rule states that a cyclic conjugated π system with 4n+2 π electrons will be very stable. **A corollary to this rule states that a cyclic conjugated π system with 4n electrons (4, 8, 12, etc.) will be very <u>unstable</u>.** Use MO theory to explain this corollary.

36. Consider two stereoisomers of the 10-member ring shown below.

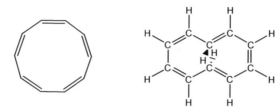

a. Construct an explanation for why the conformation on the left is very unfavorable.

b. In the conformation on the right, the two H's drawn with wedge-and-dash bonds cause the carbon backbone to flex out of planarity. Do you expect this molecule to be aromatic, anti-aromatic or non-aromatic? Explain.

37. A student states that the following molecule will be aromatic since it has 10 π electrons. What is the likely flaw in this student's reasoning?

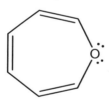

Read the assigned pages in the text, and do the assigned problems.

The Big Picture

This ChemActivity is mostly about recognizing aromatic molecules. As a result there are a large number of exotic aromatic systems shown as examples that you will likely not encounter in other contexts. Most every aromatic molecule we will encounter in this course contains a six-member ring of carbons with three double bonds, a benzene ring (also called a phenyl ring).

The next few activities explore reactions involving aromatic rings. Almost all the examples in these upcoming activities will involve benzene derivatives, but the reactions also apply to larger aromatic molecules and in certain special instances work with heterocyclic aromatic molecules.

The reactions in the next few ChemActivities are really quite amazing since the key property of benzene and other aromatic molecules is their relative nonreactivity. In fact, the reaction showcased in the next few activities were discovered accidentally by chemists who were using benzene as a solvent, assuming it was inert (which it usually is).

Common Points of Confusion

- There are many ways to represent benzene. Unfortunately, the two most common ways (a hexagon with a circle in it or using only one of the two resonance structures) are both a bit misleading.

    - Look out for the hexagon with a circle in it representation, and remember that the circle in the middle should really be a dotted-line to properly show that it represents a composite of both of benzene's two equivalent resonance structures. Beyond the fact that it is very fast and easy to draw, the one good thing about this representation is that it emphasizes the symmetrical nature of benzene.

    - The hardest part of using on resonance structure of benzene as a representation of benzene is remembering that benzene is actually symmetrical. Each of the carbon-carbon bonds in benzene is halfway between a single bond and a double bond (a 1-½ bond).

- Much student confusion revolves around anti-aromaticity. A key thing to focus on is that anti-aromaticiy is a rarely encountered phenomenon. One misconception is that an anti-aromatic molecule must violate all the rules of aromaticity. In fact, **anti-aromatic** molecule must OBEY all the rules of aromaticity EXCEPT the 4n + 2 rule. (It must instead have 4n π electrons.)

- Students sometimes get confused between the terms anti-aromatic and **non-aromatic**. They are very different. Most molecules are non-aromatic, including all noncyclic molecules. Anti-aromatic molecules turn out to be very rare because it is very easy for most cyclic molecules to violate the planarity required for anti-aromaticity by adopting a nonplanar conformation. This makes them non-aromatic instead of anti-aromatic. In fact, most examples of anti-aromatic molecules contain a **cyclobutadiene** group. This is because, unlike larger rings, the four-atom ring of cyclobutandiene is unable to twist out of a planar conformation. (Model sets show this well if you are able to make a model of a four-member ring.)

- A lone pair on an aromatic ring resides in a p orbital only if doing so makes the molecule aromatic. This is because there is an energy "cost" of putting a lone pair in a p orbital vs. a hybrid orbital due to the fact that p orbitals are closer to other orbitals than hybrid orbitals (which are, by definition, maximally spread out).

Notes

ChemActivity 2: Introduction to EAS

(In an EAS reaction involving a substituted benzene, where on the ring will the electrophile end up?)

Model 1: Electrophilic Aromatic Substitution (EAS)

When a benzene ring is treated a strong electrophile (E^+), one H is replaced by E.

The group —Z, above (emanating from no particular carbon) represents an unspecified group occupying an unspecified position on the ring.

It turns out that the identity and position of this group (Z) determines which H is replaced by E.

The mechanism of this reaction is as follows…

Construct Your Understanding Questions (to do in class)

1. In an EAS reaction, is the π bond of the aromatic ring acting as a **nucleophile** <u>or</u> **electrophile**? (Circle one and explain your reasoning.)

2. Which do you expect to be faster, an EAS reaction in which the "**Z**" group is an **Electron-Withdrawing Group (EWG)** <u>or</u> an **Electron-Donating Group (EDG)**? Explain your reasoning.

Model 2: Electron-Withdrawing and Electron-Donating Groups

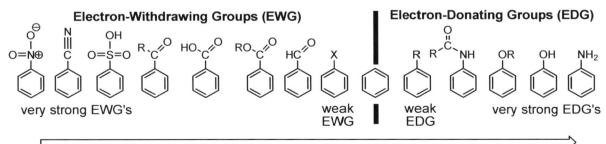

Construct Your Understanding Questions (to do in class)

3. (Check your work) Is your answer to the previous question consistent with the fact that an EAS reaction involving nitrobenzene (C_6H_5—NO_2) is <u>much slower</u> than an EAS reaction involving benzene (C_6H_6)?

4. The arrow at the bottom of Model 2 can represent the relative speed of an EAS reaction. Write "Slow EAS" at the appropriate end of the arrow and "Fast EAS" at the other end.

5. (Check your work) Write "Deactivated Rings" above the EWG's and "Activated Rings" above the EDG's, and check to make sure this is consistent with your answer to the previous question.

6. Answering the following questions will help you recognize EWG's and EDG's.

    a. Circle <u>each π bond</u>, and add <u>each lone pair</u> to the structures in Model 2.

    b. Confirm that each electron-withdrawing group (EWG) listed (except for the weakest EWG) has a π bond next to the aromatic ring.

    c. Confirm that each electron-donating group (EDG) listed (except the weakest EDG) has a lone pair on the atom next to the aromatic ring.

Model 3: Electrophilic Aromatic Substitution with Toluene

Memorization Task 2.1: ortho, meta, para nomeclature for substituents on a benzene ring

Two groups attached to <u>adjacent</u> benzene-ring carbons (e.g. C_1 and C_2) are **ortho** to each other.
Two groups attached <u>one</u> carbon apart on a benzene ring (e.g. C_1 and C_3) are **meta** to each other.
Two groups attached <u>two</u> carbons apart on a benzene ring (e.g. C_1 and C_4) are **para** to each other.

Construct Your Understanding Questions (to do in class)

7. Label the three different nitrotoluene <u>products</u> in Model 3 using the names *ortho*, *meta*, and *para* according to the definitions above.

8. Add curved arrows showing the movement of electrons in at least one of the reactions in Model 3.

    a. For each carbocation, draw the other two important resonance structures (r.s.).

    b. The *ortho* and *para* intermediates in Model 3 each have one resonance structure that stands out as being more favorable (lower in energy). Find and circle it.

    c. Construct an explanation for why the *ortho* and *para* products are formed preferentially in the reaction in Model 3.

9. The following energy diagram shows pathways to the **ortho, meta**, and **para** products.

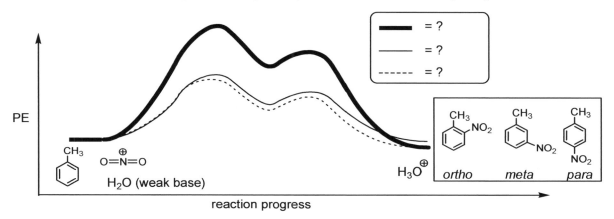

    a. Label the pathway corresponding to the **meta** product, and explain your reasoning.

    b. When toluene is mixed with the electrophile $^{+}NO_2$ only two of the three products in the box above are formed. Circle these two, cross out the one that is not observed in significant quantities.

    c. Look closely at the *ortho* and *para* intermediates (or products). One is slightly higher in potential energy due to steric effects. Which one? *ortho* or *para* [circle one].

    d. In an EAS reaction starting with toluene, the CH_3 group is said to be functioning as an "**ortho** and **para** director." Construct an explanation for this terminology.

    e. In this course we will also encounter "**meta directors**." Check the table of reagents on the next page and draw the major product of the reaction at right if the "Z" group is a *m* director such as nitro (NO_2).

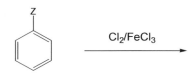

Synthetic Transformations 2.1-2.6: EAS Reagents

Reactant	Synth. Transf.	Electrophile (E^+)	Reagents	Product/s
(benzene when Z = H) *This notation indicates that an unspecified group may be attached to the ring at one of the six positions.*	2.1	$D^{\oplus}$	Any deuterated strong acid	(ring with D, Z)
	2.2	$O = N^{\oplus} = O$ (nitronium)	H_2SO_4 and HNO_3	(ring with NO_2, Z)
	2.3	$HO-\overset{\oplus}{S}(=O)=O$	anhydrous sulfuric acid (H_2SO_4 with no water)	(ring with SO_3H, Z)
	2.4	$\overset{\oplus}{Br}$ or $\overset{\oplus}{Cl}$	Br_2 and $FeBr_3$ or Cl_2 and $FeCl_3$	(ring with Br, Z) or (ring with Cl, Z)
	2.5	$R^{\oplus}$	R-X, AlX_3 (X = Cl or Br)	(ring with R, Z) R = alkyl group
	2.6	$\overset{\oplus}{C}(=O)R$	$X-C(=O)-R$ AlX_3 (X = Cl or Br)	(ring with C(=O)R, Z) R = alkyl group

---

Memorization Task 2.2: Activators = *o/p* Directors; Deactivators (usually) = *m* Directors

For reasons we will explore throughout the following three activities:

- all **activators** (electron-donating groups) are *ortho* and *para (o/p)* directors
- all **deactivators** (electron-withdrawing groups) <u>EXCEPT HALOGENS</u> are *meta (m)* directors

Halogens are the exception (again). They are the only deactivators that are o/p directors. We will explore why later.

Mem. Task 2.3: A ring with a strong EWG will not undergo a Friedel-Crafts reaction

Synthetic Transformations 2.5 and 2.6 are called "**Friedel-Crafts alkylation**" and "**Friedel-Crafts acylation**" after their inventors, Charles Friedel and James Crafts. (An **acyl group** is an RC=O group).

For reasons we will not discuss, these reactions are particularly sensitive to the rate reduction caused by a strong electron-withdrawing group, and do not yield good results when one is present.

Extend Your Understanding Questions (to do in or out of class)

10. Add the headings "***o/p* directors**" and "***m* directors**" to appropriate sides of the chart in Model 2.

11. On the table above, label Synthetic Transformations 2.5 and 2.6 "**Friedel-Crafts alkylation**" and "**acylation**," respectively, and make a note that neither works with a strongly deactivated ring

Model 4: Steric vs. Electronic Effects in EAS

For the reaction below, chemists say the ***meta*** **product is NOT formed because of unfavorable electronic effects** (because the *ortho* and *para* carbocation intermediates have more tertiary character).

Though electronic effects (e.g. carbocation stability) are generally stronger than steric effects, **steric effects** *can* have an impact.

For example: If the R group at right is large enough *para* is favored over *ortho*.

$R = -CH(CH_3)_2$ ortho (13 %)	meta (<2 %)	para (85 %)
$R = -CH_3$ ortho (63 %)	meta (<2 %)	para (35 %)

Extend Your Understanding Questions (to do in or out of class)

12. Explain the very strong preference for the *para* product when R = isopropyl (as compared to when R = methyl).

13. Explain the statement: "Without steric effects, one would expect the *ortho* product of this reaction to be exactly twice as abundant as the *para* product." (If you are stuck, *see Hint* below.)

ortho (66 %)   meta (~0 %)   para (33 %)

Expected Ratio if there were no Steric Effects

R = alkyl group

Hint: What is the difference between this product and the *ortho* structure in the box at left?

Confirm Your Understanding Questions (to do at home)

14. Show the mechanism and most likely products that result from the following reactants. (Note: Two weak bases, water and bisulfate ion also are in solution.)

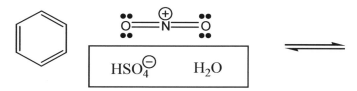

15. Sulfuric acid with no water in it is called **fuming sulfuric acid** and contains small amounts of the powerful electrophile HSO_3^+ (structure shown below). Construct a mechanism for the following reaction.

16. Draw all possible resonance structures for the carbocation intermediate in Model 3.

17. $^+NO_2$ is formed when nitric acid (HNO_3) and sulfuric acid (H_2SO_4) are mixed. Draw the Lewis structure of each and construct a mechanism that explains formation of $^+NO_2$. *Hint*: Water and HSO_4^- are the other products formed in this reaction.

18. When toluene is treated with sulfuric and nitric acids under special conditions, three nitro (NO_2) groups are substituted for hydrogens at the 2, 4 and 6 positions on the ring (the next section discusses why the 2, 4, and 6 positions are substituted). The product is a highly explosive substance called 2,4,6-trinitrotoluene. This substance is commonly known by a three letter name. What is it?

19. Construct a reasonable mechanism for the following reaction called a Friedel-Crafts alkylation. You may use curved arrows to show the formation of the carbocation electrophile in the box, or show the electrophile attached to the catalyst.

20. Draw <u>the mechanism</u> and most likely product(s) that result from the following EAS reaction called a Friedel-Crafts acylation. You may use curved arrows to show the formation of the carbocation electrophile in the box, or show the electrophile attached to the catalyst.

assume these species are present

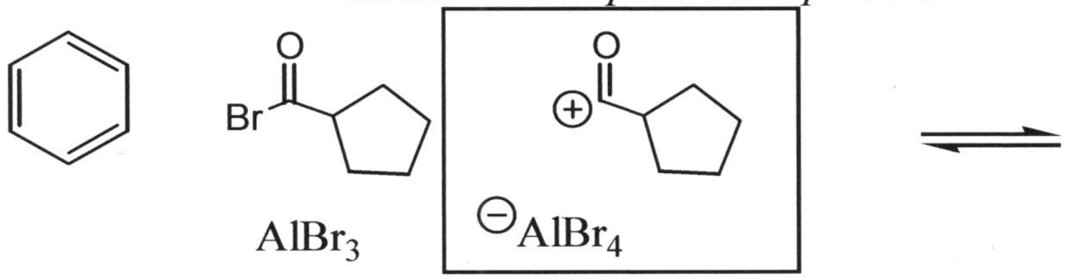

21. Consider the reactions below:

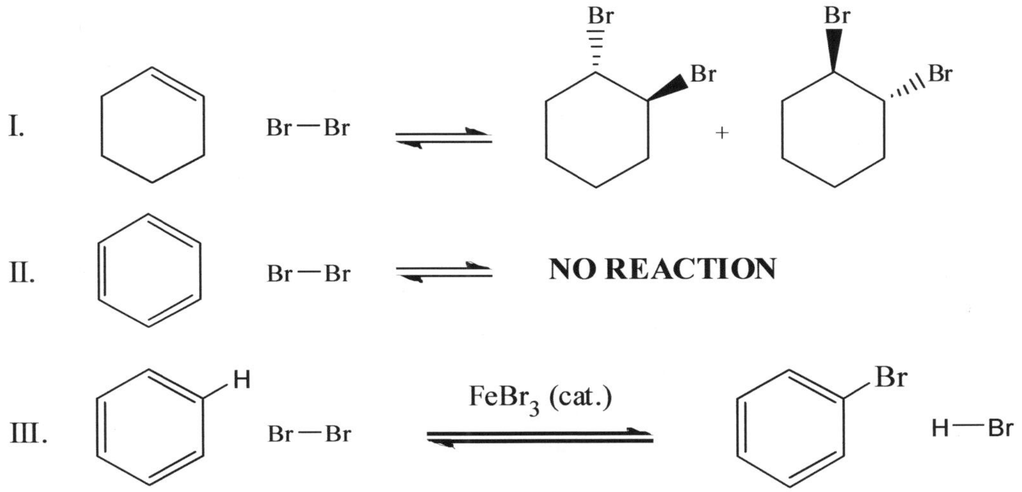

I.

II. **NO REACTION**

III.   FeBr$_3$ (cat.)

a. Why does Br$_2$ react with cyclohexene but not with benzene? (*See* Rxns I & II, above)

b. Construct a mechanism for Reaction III that shows FeBr$_4$ acting as a base in the last step, generating FeBr$_4$H, which decomposes into HBr and FeBr$_3$, regenerating the Lewis acid catalyst.

22. Recall that an alkyl group (R) is slightly electron-donating. This means it donates electron density into an attached aromatic ring. Based on this information, which of the following EAS reactions do you expect to be faster? Explain your reasoning.

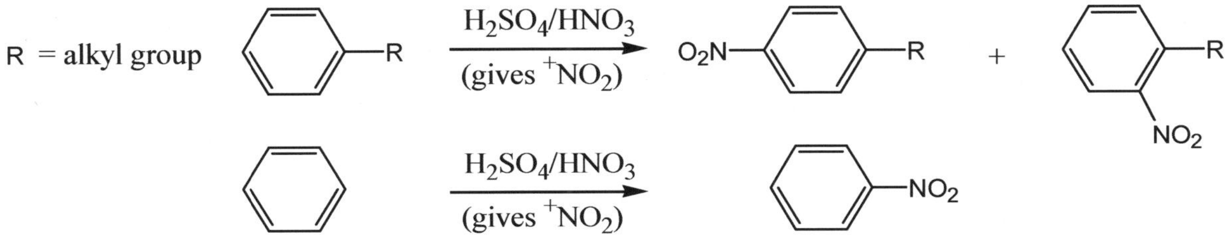

R = alkyl group   H$_2$SO$_4$/HNO$_3$
(gives $^+$NO$_2$)

H$_2$SO$_4$/HNO$_3$
(gives $^+$NO$_2$)

23. Draw an example of an aromatic molecule that…

    a. will NOT undergo either Synthetic Transformation 2.5 or 2.6.

    b. *will* undergo Synthetic Transformations 2.5 or 2.6.

24. Of the choices in brackets, circle the word or phrase that makes the sentence true.

    a. The pi system of the ring acts as [**a nucleophile** *or* **an electrophile**] in an EAS reaction.

    b. The [**more** *or* **less**] electron-rich the pi system of the aromatic ring, the faster the rate of EAS.

25. A nitro group is a very powerful electron-withdrawing group. Which do you expect will undergo EAS reaction faster: **benzene** <u>or</u> **nitrobenzene** (circle one and explain your reasoning).

26. Construct an explanation for the following finding: Even with the best electrophile (E^+), di-nitro benzene undergoes EAS extremely slowly, so slowly that only trace amounts of product are observed under normal conditions.

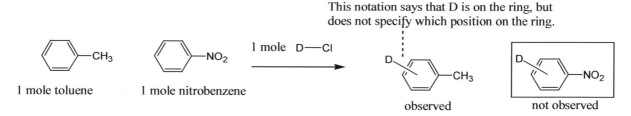

27. When a flask containing a mixture of one mole of nitrobenzene and one mole of toluene is treated with one mole of D-Cl, deuterium is incorporated into the toluene ring, but not the nitrobenzene ring. Explain. That is, why does the toluene "hog" all the D-Cl while the nitrobenzene does not get any?

This notation says that D is on the ring, but does not specify which position on the ring.

1 mole toluene   1 mole nitrobenzene   1 mole D—Cl

observed   not observed

Read the assigned pages in your text and do the assigned problems.

The Big Picture

This is the first of several activities on the topic of electrophilic aromatic substitution (EAS). There are two key concepts (and a bunch of new synthetic transformations to memorize) in this activity. The concepts are: 1) the more electron rich the ring, the faster the EAS reaction, and 2) a group already on the ring will determine the placement of a second group add to the ring via EAS.

This activity provides only a brief introduction to the latter, and explores only the argument for why alkyl groups direct *ortho* and *para*. Essentially this is a carbocation intermediate stability argument, with the carbocations leading to *ortho* and *para* being more favorable than the one leading to *meta*.

The next activity explores directing effects based on resonance arguments. All of these arguments are important and provide the foundation for more complex arguments later in the course, and it is essential that you understand them; however, you should also know the simple rule that all electron donating groups are *o/p* directors, and all electron withdrawing groups (except halogens) are *m* directors.

Common Points of Confusion

- Students get confused and assume that because a nitro group is a *meta* director, it must therefore add *meta* to whatever group is already on the ring. This represents flawed logic. Directing effects are a function of the group(s) ALREADY ON THE RING. For example, when adding a nitro group to toluene, it is irrelevant that nitro is a *m* director. The nitro group will add *ortho* or *para* because the CH_3 group already on the ring is an *o/p* director.

- Students often confuse Friedel-Crafts reagents and EAS halogenation reagents. This happens especially often with the reagents CH_3Br and CH_3Cl. These reagents do NOT DELIVER A BROMINE/CHLORINE ATOM; they deliver a methyl group. Br_2 or Cl_2 (not an alkyl halide!) is used (in conjunction with an appropriate Lewis acid catalyst) to replace an H with a halogen on an aromatic ring.

- Students frequently make the mistake of assuming that EAS halogenations do not work with a strong deactivator (such as NO_2) attached to the ring. This is incorrect: **You can halogenate nitrobenzene or other rings with one strong deactivator.** The logic behind this error seems to be that such halogenations employ a Lewis acid catalyst FeX_3 that is similar to the Friedel-Crafts catalyst (AlX_3). It turns out that the halogenations are less sensitive to deactivation than the alkylations and acylations.

- No ordinary EAS reactions work with very deactivated rings. As a rule of thumb, a ring is considered very deactivated if it has two resonance electron-withdrawing groups (and no resonance electron-donating groups to counterbalance this).

Notes

ChemActivity 3: EAS Resonance Effects

(How do resonance-withdrawing and donating groups affect rate and placement in EAS?)

Model 1: Second-order Resonance Structures

For this activity we define the term **second order resonance structure** to mean any resonance structure in which a π bond has been broken to generate two extra formal charges.

Think of second-order resonance structures as "assistant resonance structures." They are not as important as regular (first-order) resonance structures, but can help us make predictions in certain instances.

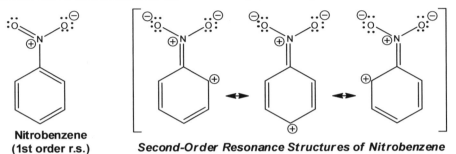

Nitrobenzene
(1st order r.s.)

Second-Order Resonance Structures of Nitrobenzene

Construct Your Understanding Questions (to do in class)

1. Label each first-order and each second-order resonance structure of acetone.

    acetone

2. In Model 1, add curved arrows to nitrobenzene showing how to generate one of the second-order resonance structures. Then use curved arrows to generate each of the subsequent second-order resonance structures.

3. The second-order resonance structures of nitrobenzene tell us that the nitro group generates partial positive charge at certain carbons on the aromatic ring. Complete the composite drawing of nitrobenzene below by placing a δ+ on appropriate carbons in the ring.

    Composite Drawing of Nitrobenzene
    (a combination of all first- and second-order resonance structures)

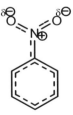

4. (*Review*) In the first step of an electrophilic aromatic substitution (EAS) reaction **the aromatic ring is acting as a nucleophile** and making a bond with the electrophile (E^+).

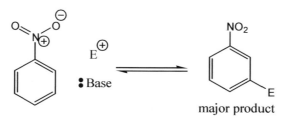

a. Which is more likely to make a bond with an electrophile, **a carbon that is electron-rich, or a carbon that is electron-poor and holds a δ+ charge** (circle one)?

b. Assume that Z (above) is a nitro group and add δ+ charges to appropriate ring carbons.

c. Construct an explanation for why this reaction yields *meta* **as the major product**.

Memorization Task 3.1: Resonance vs. Inductive Withdrawing/Donating Groups

Inductive electron-withdrawing (or donating) group = group that withdraws or donates electrons only via inductive effects. This is usually a **weak** effect.

   Example: alkyl group (*see* ChemActivity 2)

Resonance electron-withdrawing (or donating) group = group that withdraws or donates electrons via π bonds as demonstrated by 2nd order resonance structures—this is usually a **STRONG** effect.

   Examples: (withdrawing) nitro group or any other group with a π bond next to the ring
                (donating) amino group or any other group with a lone pair next to the ring

---

5. Draw second-order resonance structures of the aromatic starting material (benzaldehyde) and use these to justify why the EAS reaction in the box yields mostly *meta*-bromobenzaldehyde.

(Note that a more complete explanation for why the meta product is preferred must address the relative energies of the carbocation intermediates. Questions that focus on this issue can be found in the homework for today's ChemActivity.)

6. Draw examples of two mono-substituted benzene rings (other than nitrobenzene and benzaldehyde) that also are expected to preferentially direct an electrophile to the *meta* position. (*See* Model 2 from the previous ChemActivity.)

7. Add curved arrows to aniline to generate the second-order resonance structure shown. Then use curved arrows to generate the two missing second-order resonance structures.

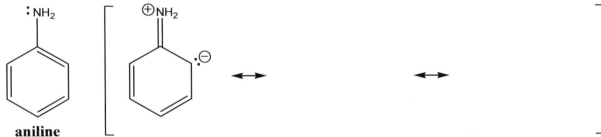

aniline   *Second-Order Resonance Structures for Aniline*

8. Place a δ– on appropriate carbons in the ring at right to complete a composite drawing of aniline showing a combination of all first and second order resonance structures.

9. Recall the first step of an EAS reaction.

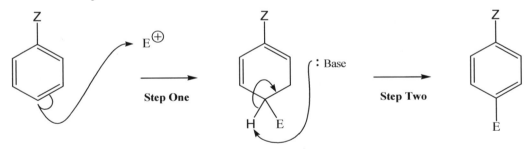

Step One   : Base   **Step Two**

a. Assume that Z (above) is an amino group (NH₂), and add a δ– to appropriate ring carbons on the structure of the starting material.

b. Construct an explanation for why, when Z = NH₂, the major products are *ortho* and *para*.

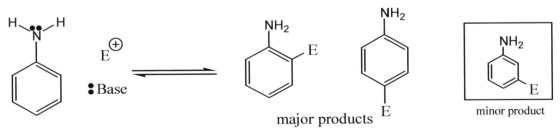

: Base   **major products**   minor product

Extend Your Understanding Questions (to do in or out of class)

10. Draw second-order resonance structures of the aromatic starting material (phenol) and use these to justify why the EAS reaction in the box yields mostly the *ortho* and *para* products.

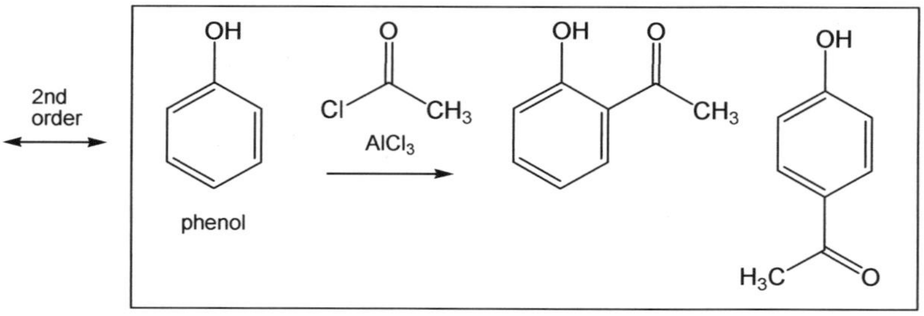

(Note that a more complete explanation for why the ortho and para products are preferred must address the relative energies of the carbocation intermediates. Questions that focus on this issue can be found in the homework for this ChemActivity.)

11. Draw examples of two mono-substituted benzene rings (other than aniline and phenol) that also are expected to preferentially direct an electrophile to the *ortho* and *para* positions.

12. As stated earlier, resonance electron-withdrawing and donating effects are usually strong in comparison to other effects (e.g. steric or inductive effects). Circle the answer that completes the sentence…

    a. A resonance electron-withdrawing group is a strong ***ortho/para* director** <u>or</u> ***meta* director**.

    b. A resonance electron-donating group is a strong ***ortho/para* director** <u>or</u> ***meta* director**.

13. Which identity of Z is expected to give a higher yield of the di-chloro and tri-chloro products shown below: **Z = CH₃ <u>or</u> Z = OH**? (Circle one and explain your reasoning.)

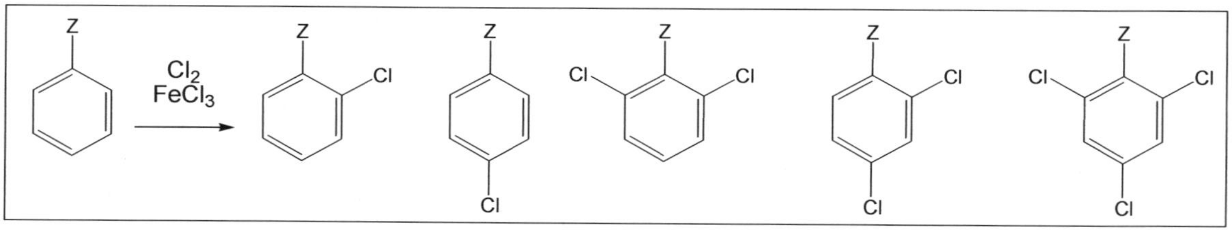

Model 2: Polysubstitution in Friedel-Crafts Alkylations

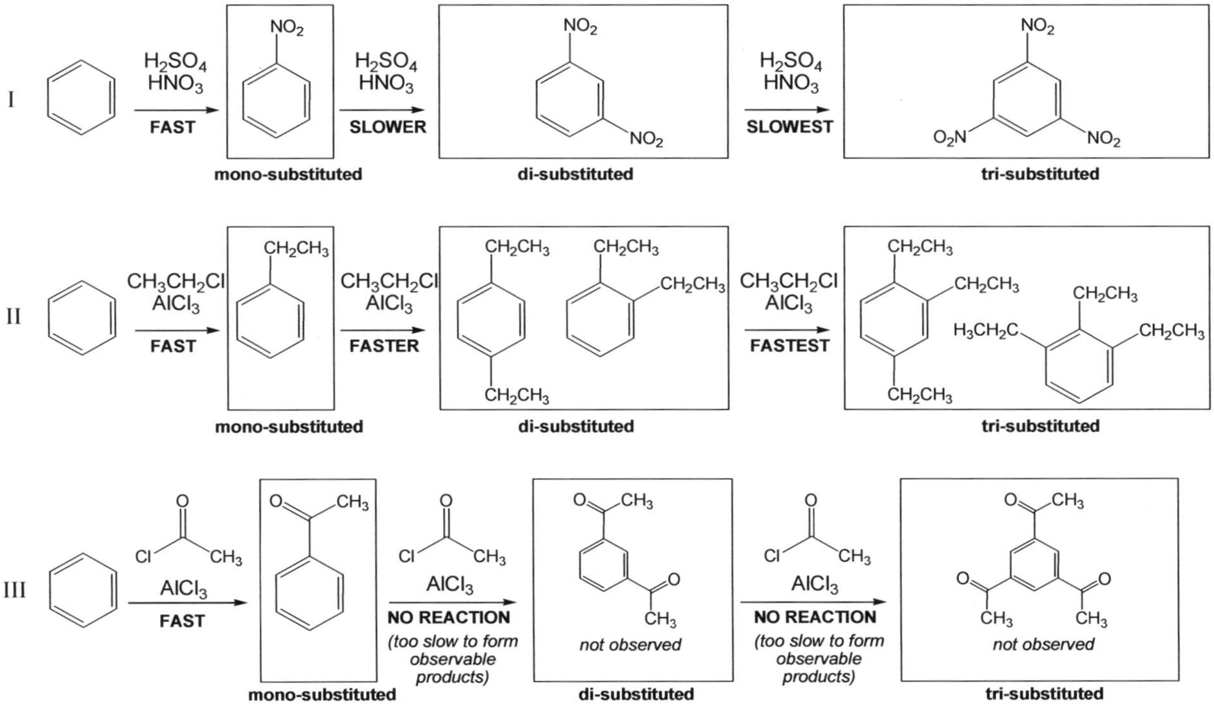

Extend Your Understanding Questions (to do in or out of class)

14. For **Reaction Sequence II** in Model 2, if you mix equal amounts of benzene and electrophile (e.g. $CH_3CH_2Cl + AlCl_3$ catalyst) the result is a large amount of unreacted starting material and the di- and tri-ethyl products shown in **Reaction Sequence II** (and almost none of the mono-ethyl product, ethylbenzene). Explain.

15. In contrast, for **Reaction Sequence III**, if you mix equal amounts of benzene and electrophile you will isolate only one of the products shown in **Reaction Sequence III**. Circle the single product formed and <u>explain why this result is so different from Reaction Sequence II</u>.

16. (Check your work) Are your answers above consistent with Memorization Task 3.2?

Memorization Task 3.2: Multiple EAS Substitutions with Highly Activated Ring

EAS reactions in which the ring has one or more strong electron-donating groups (= resonance electron-donating groups) often result in products with more than one electrophile on the ring. Note that this is <u>especially true for Friedel-Crafts alkylations, (but *not* for Friedel-Crafts acylations!!).</u>

Model 3: Carbocation Rearrangement in Friedel-Crafts Alkylation

The electrophile in a Friedel-Crafts alkylation is essentially a carbocation. (Note that due to stabilization by $^-AlX_4$ it is acceptable to propose a + charge on a methyl or primary carbon in the intermediate of a Friedel-Crafts alkylation).

Like ordinary carbocations, <u>Friedel-Crafts alkylation electrophiles will undergo favorable rearrangements</u>.

Extend Your Understanding Questions (to do in or out of class)

17. Use curved arrows to show the mechanism of formation of the major mono-alkylation product. Show the electrophile as an ordinary carbocation. (i.e., no need to show the $^-AlCl_4$ counterion.)

18. For each, <u>assume carbocation rearrangement</u> and draw a likely F-C <u>mono</u>-alkylation product. *(Assume that benzene is in huge excess so that polyalkylation products do not dominate.)*

Memorization Task 3.3: F-C acylation not plagued by rearrangement

The electrophile in F-C acylation is stable with respect to rearrangement due to resonance stabilization.

Synthetic Transformation 3.1: Reduction of C=O (and/or C=C)

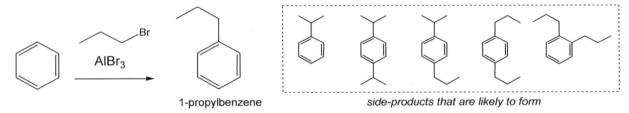

Confirm Your Understanding Questions (to do at home)

19. Based on Synthetic Transformation 3.1, summarize the differences between the effect of [H_2/Palladium metal] versus [Zn in mercury amalgam with HCl] on C=O or C=C bonds next to a benzene ring, and not next to a benzene ring.

20. The following synthesis of 1-propylbenzene gives multiple side products. Propose an alternate synthesis from benzene that gives a high yield of the desired product with little or no side products.

1-propylbenzene   side-products that are likely to form

21. Draw as many different constitutional isomers as you can of a mono-substituted benzene ring in which the single subsituent has a molecular formula C_2H_4NO and includes a carbonyl group (C=O). [There are **8** possibilities, of which 3 are resonance electron-donating groups, and 2 are resonance electron-withdrawing groups.]

    a.   Label the isomers that are resonance electron-donating groups.

    b.   Label the isomers that are resonance electron-withdrawing groups.

    c.   For each, identify whether the isomer's substituent is a strong *o/p* director, strong *m* director, weak *o/p* director, or weak *m* director.

22. Draw three second-order resonance structures for phenol.

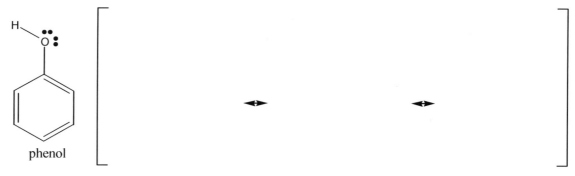

a.   Explain why the second-order resonance contributors are less important than first-order resonance structures, and contribute only a small amount to our overall understanding of phenol.

b.   Explain the following statement: The second-order resonance structures help explain why substituents with a lone pair (such as –OH) activate the *ortho* and *para* positions toward electrophilic aromatic substitution (EAS).

23. Draw three second-order resonance structures for benzoic acid.

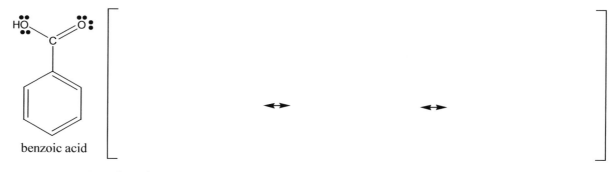

a.   Based on these second-order resonance structures, do you expect the carboxylic acid group (COOH) to be a resonance-donating group or a resonance-withdrawing group?

b.   Explain the following statement: The second-order resonance structures help explain why a carboxylic acid group **deactivates** the *ortho* and *para* positions toward electrophilic aromatic substitution (EAS).

24. Second-order resonance structures can be used to predict the regiochemistry of an EAS reaction. An identical, but more rigorous prediction comes from the heights of the activation barriers. According to the Hammond Postulate, the relative energies of the intermediates can be used to approximate the relative heights of these activation barriers. Simply put... **The pathway with the most favorable carbocation intermediate is expected to dominate.**

Consider electrophilic aromatic substitution (EAS) performed on nitrobenzene.

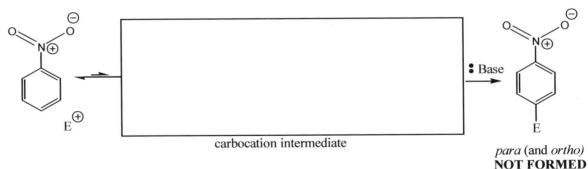

major product

minor products

a. How does the placement of E^+ on the nitrobenzene ring differ from an EAS reaction starting with toluene or aminobenzene (aniline)?

b. Draw the intermediate on the reaction pathway to the major (*meta*) product. (Be sure to include **all important resonance structures**.)

carbocation intermediate

meta
ONLY product

c. Draw the intermediate on the reaction pathway to the *para* product. (Be sure to include **all important resonance structures**.)

carbocation intermediate

para (and *ortho*)
NOT FORMED

d. Any resonance structure in which two + charges are next to each other is very unfavorable. Circle all unfavorable resonance structures above.

e. Construct an explanation for why the intermediate on the reaction pathway to the *meta* product is lowest in potential energy.

f. Draw an energy diagram showing all three pathways (*ortho, meta* and *para*).

g. Construct an explanation for why the *meta* product is strongly favored in this reaction over the *para* (and *ortho*) product.

25. Aniline gives only *ortho* and *para* products in an EAS reaction.

    a. Draw the intermediate on the reaction pathway to the *para* product. Be sure to draw all **FOUR** important resonance structures.

aniline

    b. Draw the intermediate on the pathway to the *meta* product.

    c. Construct an explanation for why the intermediate on the pathway to the *meta* product is higher in potential energy than the intermediate on the pathway to the *para* product.

    d. Draw an energy diagram showing all three pathways (*ortho, meta* and *para*).

26. Aniline reacts with a given electrophile 100 times faster than toluene and 1000 times faster than benzene.

    a. Explain why an EAS intermediate for aniline such as the one in part a) above is lower in potential energy than the intermediate you drew for *para*-substitution of toluene.

    b. Construct an explanation for why aniline undergoes EAS much faster than toluene.

27. If you add acid to aniline you change the amine group (NH_2) into an ammonium ion group ($^+NH_3$). Draw the intermediate carbocations on the pathways to the *meta* and *para* products, respectively; predict whether an ammonium ion group is an *o/p* director or a *m* director; and explain your reasoning.

28. Draw a mechanism to explain the formation of each of the two Friedel-Crafts products.

29. Give an example (not appearing in this ChemActivity) of an alkyl halide (R–X) that will

    a.   likely undergo rearrangement during a Friedel-Crafts alkylation.

    b.   NOT undergo rearrangement during a Friedel-Crafts alkylation.

30. Shown below are two ways of making the same target product starting from benzene. Synthetic pathway b gives a higher % yield of the desired product. Explain why.

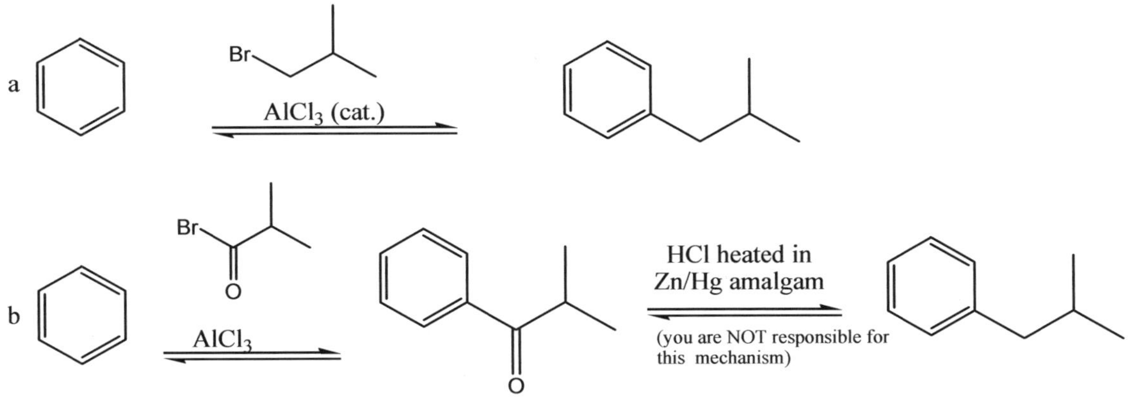

31. What reagents could you use to carry out the following reactions?

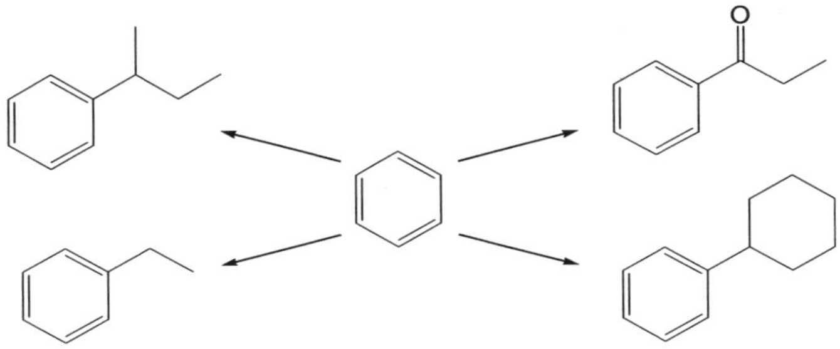

32. If you mix one mole of benzene + one mole of sulfuric acid + one mole of nitric acid you will isolate only one product. What is this single product? Explain why only this product is formed.

33. Explain why Reaction Sequence II in Model 2 yields a large amount of unreacted benzene starting material when equimolar portions of benzene and electrophile are mixed.

34. Draw all important resonance structures for the following carbocation. This resonance stabilization is used to explain that acyl carbocations do not rearrange like alkyl carbocations.

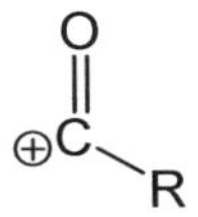

35. Using a huge excess of benzene in a F-C alkylation will reduce the likelihood that polyalkylation products will form. Construct an explanation for why.

36. Predict the major product if each of the following molecules are treated with excess Zn(Hg)/HCl. Then draw the products that result if they are treated with excess H_2 and palladium.

37. Draw the mono-substitution product (if any) that results when toluene is treated with $AlCl_3$ and each of the following halides (assume a large excess of toluene).

38. Consider a solution consisting of one molar equivalent of each of the following: benzene, nitrobenzene, and dinitrobenzene. If you treat this mixture with one molar equivalent of an electrophile (e.g. O_2N^+ made from H_2SO_4/HNO_3), what product is likely to dominate the final product mixture? That is, which starting material, **benzene, nitrobenzene, or dinitrobenzene** is most likely to react with the $^+NO_2$ electrophile? (Circle one and explain your reasoning.)

39. Consider a hypothetical solution consisting of one molar equivalent of each of the following: benzene, ethylbenzene, and diethylbenzene. If you treat this mixture with one molar equivalent of electrophile (e.g. R^+ made from R—$Cl/AlCl_3$), what product is likely to dominate the final product mixture? That is, which ring (**benzene, ethylbenzene, or diethylbenzene**) is most likely to react with the R^+ electrophile? (Circle one and explain your reasoning.)

Read the assigned pages in your text and do the assigned problems.

The Big Picture

This activity continues our exploration of electrophilic aromatic substitution (EAS) by highlighting the directing effects of resonance electron donating and withdrawing group. The second order resonance structure arguments made in this activity are very simple, and give the correct answers, but the intermediate energy arguments in the homework are generally considered more robust.

The next activity concludes our exploration of EAS by investigating competing directing effects.

Common Points of Confusion

- Friedel-Crafts reactions are very sensitive, and are only useful in certain circumstances. Watch out for the following problems: 1) F-C reactions do not work with a strongly deactivated ring, 2) multiple alkylations can occur, and 3) carbocation rearrangements can occur.

- Students notice that *meta* directors very often have a carbonyl group. Be careful with this, especially for amides and esters. The group is a *meta* director only if the carbonyl is directly adjacent to the aromatic ring. Very often, one constitutional isomer will be a *meta* director and another an *o/p* director.

Notes

ChemActivity 4: EAS Competing Effects

(What happens if you have two substituents on a ring that do not agree in terms of directing effects?)

Model 1: Inductive Effects vs. Resonance Effects

Recall that an **inductive effect** is the donation or withdrawal of electron density through polarized σ bonds (as shown below, left).

A **resonance effect** is the donation or withdrawal of electron density through π bonds as demonstrated by first- or second-order resonance structures (as shown below, right).

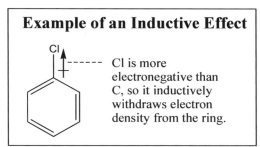

Example of an Inductive Effect

Cl is more electronegative than C, so it inductively withdraws electron density from the ring.

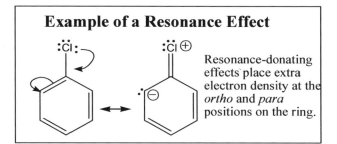

Example of a Resonance Effect

Resonance-donating effects place extra electron density at the *ortho* and *para* positions on the ring.

- All **activators** (electron-donating groups) are *ortho/para* directors in EAS

- All **deactivators** (electron-withdrawing groups) <u>except halogens</u> are *meta* directors in EAS

Construct Your Understanding Questions (to do in class)

1. Halogens are net EAS deactivators. Based on this, which is stronger in the case of halogens, **inductive effects** <u>or</u> **resonance effects**? (Circle one and explain your reasoning.)

2. Let us explore why halogens are deactivators, but *o/p* directors.

   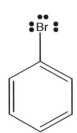

   a. Halogens are powerful inductive-withdrawing groups. This means that a halogen **removes electron density from all six carbons of the ring**. Though the size of this effect is dependent on the distance from bromine.

      Represent this at left by placing a LARGE $\delta+$ on each C of the ring.

   b. However, a halogen can resonance-donate back into the ring with one of its lone pairs (as demonstrated by second-order resonance structures).

      To represent this, on the structure of bromobenzene at left, put a small $\delta-$ next to the <u>three</u> ring carbons that receive electron donation from Br. (It may help to sketch 2nd order resonance structures of bromobenzene)

   c. Circle the <u>carbons</u> on bromobenzene that are most likely to bond to an electrophile (E^+), and explain your reasoning.

d. (Check your work) Is your answer to part c consistent with the fact that halogens are *o/p* directors?

Model 2a: EAS Reactions with Di-Substituted Rings

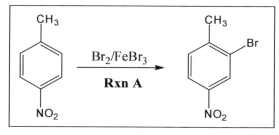

Construct Your Understanding Questions (to do in class)

3. Consider the starting material in **Rxn A** in Model 2a.

a. Mark the <u>open</u> position(s) on the ring where the CH_3 group would direct an E^+.

b. Mark the <u>open</u> position(s) on the ring where the NO_2 group would direct an E^+.

c. Are these two directing effects **opposed to one another** <u>or</u> **in agreement** (circle one)?

4. Consider the starting material in **Rxn B** in Model 2a.

a. Mark the <u>open</u> position(s) on the ring where the ethyl group (Et) would direct an E^+.

b. Mark the <u>open</u> position(s) on the ring where the OH group would direct an E^+.

5. (Check your work) In **Rxn B** the directing effects of the two substituents are <u>not</u> in agreement.

a. According to the product in **Rxn B**, which substituent dominates (wins) in terms of directing effects, **the OH group** <u>or</u> **the alkyl (R) group**? (circle one)

b. Which is a <u>stronger</u> activator, **an OH group** <u>or</u> **an alkyl (R) group**? (circle one)

c. (Check your work) Are your answers to a and b, above, consistent with one another?

Memorization Task 4.1: Two EAS activators in competition

When two EAS activators are in <u>disagreement</u> in terms of directing effects assume the <u>stronger</u> of the two activators "wins" and directs the placement of a majority of the electrophiles.

Model 2b: EAS Reactions with Di-Substituted Rings

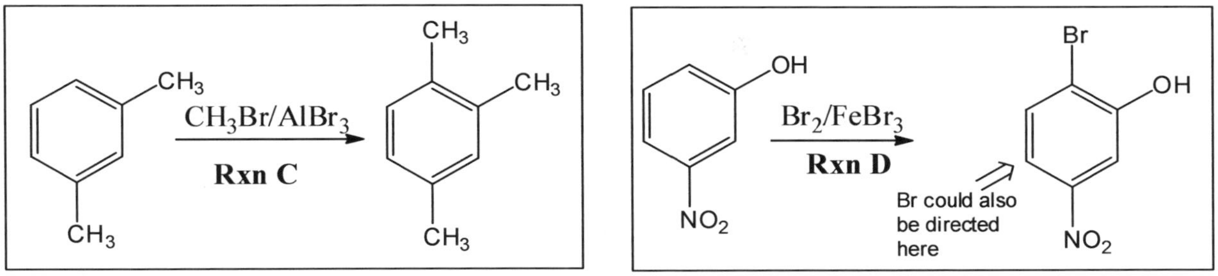

6. Consider the starting material in **Rxn C** in Model 2b.

    a. Mark the <u>open</u> position(s) on the ring where each CH_3 group might direct an E^+.

    b. Construct an explanation for why the product below is formed in very small amounts compared to the product shown in **Rxn C**.

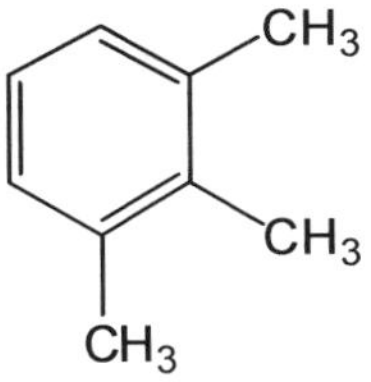

Minor Product

Memorization Task 4.2 Steric effects in EAS

Placement of an electrophile between two groups is often unfavorable due to steric effects.

---

7. Consider the starting material in **Rxn D** in Model 2b.

    a. Mark the <u>open</u> position(s) on the ring where the OH group would direct an E^+.

    b. Mark the <u>open</u> position(s) on the ring where the NO_2 group would direct an E^+.

    c. According to the major product shown in **Rxn D**, which substituent dominates (wins) in a competition of directing effects between an activator and a deactivator?

Memorization Task 4.3: EAS activator in competition with an EAS <u>de</u>activator

When an EAS activator is in disagreement in terms of directing effects with an EAS deactivator, assume the activator "wins" and directs the placement of the electrophile.

Extend Your Understanding Questions (to do in or out of class)

8. Consider an EAS starting material similar to the starting material in Rxn D, except with a *tert*-butyl alkyl group. This reaction (shown below) yields the product shown. Construct an explanation for why in Rxn D two different products are formed, one with the Br *ortho* and the other Br *para* to the *o/p* directing group (which is OH in the case of Rxn D).

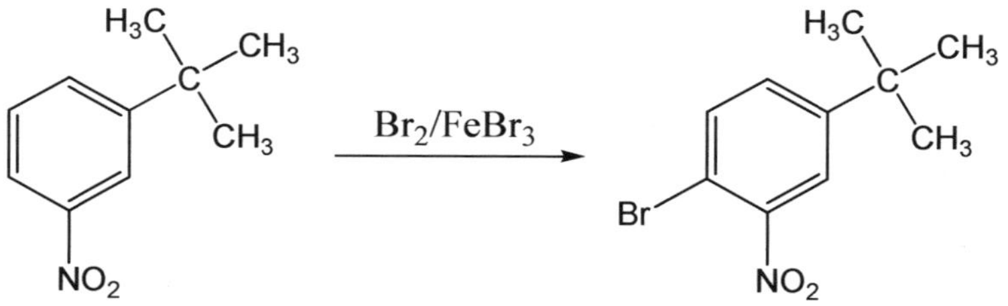

9. What is the result if you replace the reagents $Br_2/FeBr_3$ in Rxn A Model 2a and Rxn D in Model 2b with $CH_3Br/AlBr_3$?

10. (Check your work) Is your answer to the previous question consistent with Memorization Task 2.3 (found in ChemActivity 2)?

11. What is the result of the following reaction?

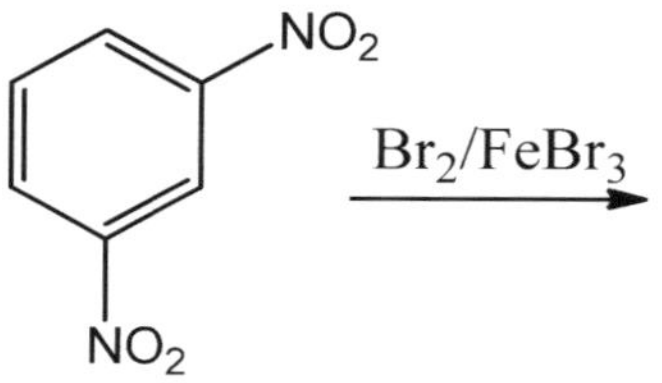

12. (Check your work) Is your answer to the previous question consistent with Memorization Task 4.4?

Memorization Task 4.4: Ring with two or more strong deactivators → very slow EAS

Only under extreme conditions will a ring with two or more strong **deactivators** undergo EAS.

Model 3: EAS Directing Effects (summary)

Identity of Z	Inductive Effects	Resonance Effects	Product Regio-chemistry	Relative Reaction Rate
Anilines —NH$_2$, —NHR, —NR$_2$	weak e⁻ **withdraw**	strong e⁻ **donation**		very very fast
Phenols —OH, —OR	moderate e⁻ **withdraw**	strong e⁻ **donation**		very fast
Alkyl groups —CH$_3$, —CH$_2$CH$_3$ etc.	weak e⁻ **donation**	—		medium fast
Benzene —H	—	—	—	1
Halogens —I, —Br, —Cl, —F	strong e⁻ **withdraw**	weak e⁻ **donation**		slow
Acyl groups	strong e⁻ **withdraw**	strong e⁻ **withdraw**		medium slow
Sulfate group	strong e⁻ **withdraw**	strong e⁻ **withdraw**		medium slow
Nitro group —NO$_2$	strong e⁻ **withdraw**	strong e⁻ **withdraw**		medium slow
Cyanide group —C≡N	strong e⁻ **withdraw**	strong e⁻ **withdraw**		medium slow
Ammonium —NR$_3$ (R = alkyl or H)	Very strong e⁻ **withdraw**	—		medium slow

Extend Your Understanding Questions (to do in or out of class)

13. Above, fill in each empty box in the Column labeled **Product Regiochemistry** with *o/p* or *m*.

14. (Check your work) Identify the one *meta* director listed that is <u>not</u> a strong resonance electron withdrawing group.

15. (Check your work) Identify the one *ortho/para* director that is not a net activator.

Confirm Your Understanding Questions (to do at home)

16. Consider the following reactions:

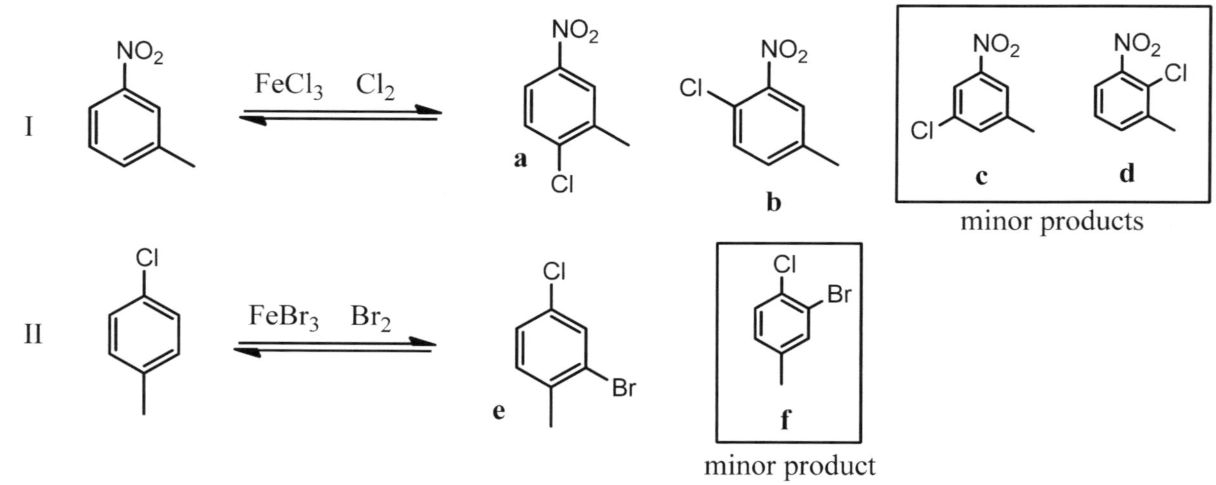

a minor products

b

f minor product

a. Construct an explanation for why Product **c** is a minor product in Rxn I.

b. Construct an explanation for why Product **d** is a minor product in Rxn I.

c. Construct an explanation for why Product **f** is a minor product in Rxn II.

17. In the reaction below, two major products are observed.

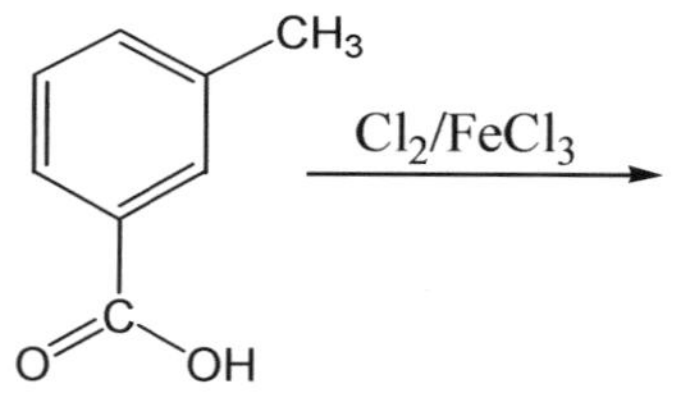

a. Draw them, and construct an explanation for why they are formed instead of the other two possibilities.

b. Which of the two major products do you expect to dominate and why?

18. Mark each of the following statements True or False based on your current understanding. (**If false**, cite an example of a substituent for which it is false.)

a. T or F: A strong activator overpowers the directing effects of a weak activator.

b. T or F: A weak activator overpowers the directing effects of a deactivator.

c. T or F: All activators are *o/p* directors.

d. T or F: All deactivators are *m* directors.

19. Circle <u>each</u> of the following that helps explain why EAS with fluorobenzene is slower than EAS with phenol (hydroxybenzene)?

      (1)   F is more electronegative than O, generating stronger inductive effects.

      (2)   F holds it lone pairs very tightly, making it a weaker pi donor than O.

      (3)   F has more lone pairs than O.

      (4)   F is smaller in size than O.

20. Consider the following reactions:

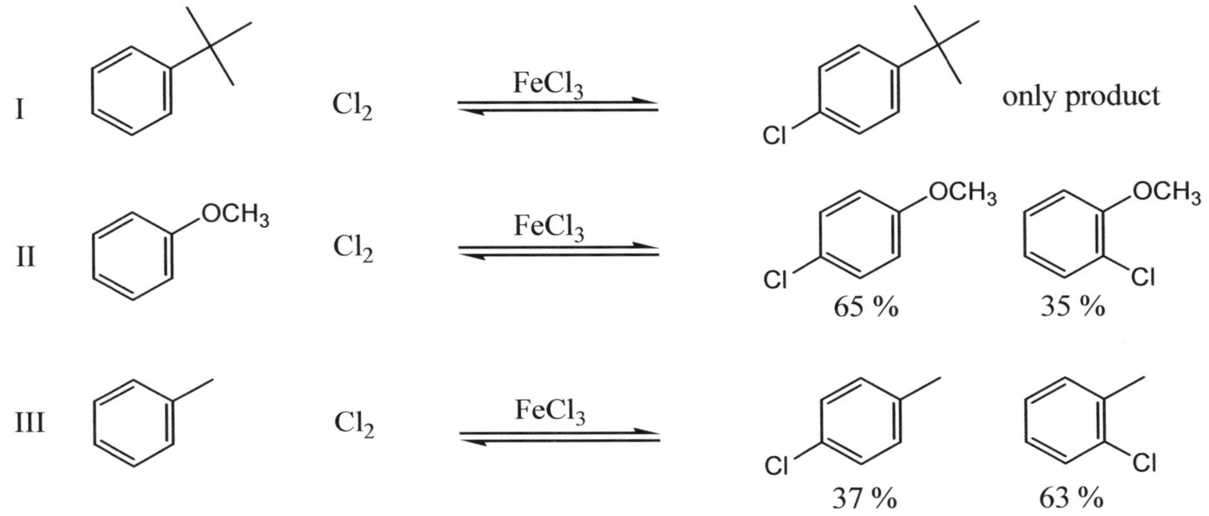

a.   Construct an explanation for why <u>only</u> the *para* product is observed in reaction I.

b.   Neglecting steric effects, the expected ratio of *para:ortho* is 1:2 for reactions I, II and III. Explain why this is the case, and why Reaction III is closest to the expected 1:2 ratio.

Read the assigned pages in your text and do the assigned problems.

The Big Picture

A key issue in this activity is the competition between directing effects. This activity gives you two rules (below) that work well in most instances. See the last two bullets in Common Points of Confusion for discussion of cases where these rules may not be sufficient.

> **Memorization Task 4.1**: When two EAS activators are in disagreement in terms of directing effects assume the <u>stronger</u> of the two activators "wins" and directs the placement of a majority of the electrophiles

> **Memorization Task 4.2**: When an EAS activator is in disagreement in terms of directing effects with an EAS deactivator, assume the <u>activator</u> "wins" and directs the placement of the electrophile

You may wonder why you were asked to memorize a number of restrictions on Friedel-Crafts reactions. The reason is that we will begin using these reactions more and more to solve synthesis problems since Friedel-Crafts reactions are one of only a few ways to attach a carbon or carbon chain to a benzene ring.

At some point in your course you will likely encounter a reaction called ***Nucelophilic*** Aromatic Substitution or NAS. In many ways, EAS and NAS are opposites. For example, in EAS the ring functions as a nucleophile (and reacts with an electrophile), while in NAS the ring functions as an electrophile (and reacts with a nucleophile). NAS is generally less useful than EAS, but the two are easily confused. For students who have covered NAS, the next ChemActivity has some questions to help you distinguish EAS from NAS, and bring the two together to showcase general strategies for synthesizing aromatic targets using EAS, NAS, and side chain reactions starting from benzene.

Common Points of Confusion

- The mechanism of EAS is fairly straightforward. Where many students run into difficulty is sorting out the many restrictions, especially those revolving around the very sensitive Friedel-Crafts reactions. The easiest way to remember all the Memorization Tasks in this book is to understand as much as you can about why each statement is true.

- Students commonly confuse EAS (*Electrophilic* Aromatic Substitution) with a complementary reaction called NAS (*Nucelophilic* Aromatic Substitution). The two biggest points of confusion between them are as follows:

  o As the name implies, in NAS a nucleophile substitutes for another group. In EAS an electrophile substitutes for an H, but in NAS the nucleophile substitutes for a halogen leaving group. Keep an eye out for that leaving group! Without it, NAS cannot occur.

  o In NAS, the benzene ring is functioning as an electrophile. This means that electron *withdrawing* groups on the benzene ring *speed* the rate of NAS, and electron donating groups slow down NAS. This is the opposite from EAS. (Electron donating groups speed EAS, and electron withdrawing groups slow down EAS.)

- Many students are left wondering, what happens when you have a strong deactivator in competition with a weak activator? Memorization Task 4.2 (reprinted above) would have you assume that the activator still wins. In turns out that the real world is messy, and that in the laboratory such a reaction will give a messy mixture of products. For example, halogenation of *meta*-nitrotoluene gives a mixture of all possible products. For the purposes of this activity (for example, in Question 16) we assume that the rules above hold true.

- What happens if you have two deactivators in competition in terms of directing effects? The answer is often moot since two deactivators (especially if they are both strong deactivators) will usually slow a reaction to the point where it is not useful for generating product.

Notes

ChemActivity 5: EAS Synthesis Workshop

(How and why would you change an *ortho/para* director into a *meta* director or vice versa?)

Model 1: Synthesis and Retrosynthesis

Often the best way to design a synthesis of a **target** molecule is to mentally pull the target apart into simpler idealized parts (called **synthons**) using a technique called **retrosynthesis.**

The target below can be made in two steps from benzene. Forward steps are shown with reaction arrows; a retrosynthetic step is shown with a special **retrosynthetic arrow** (labeled "retro" below).

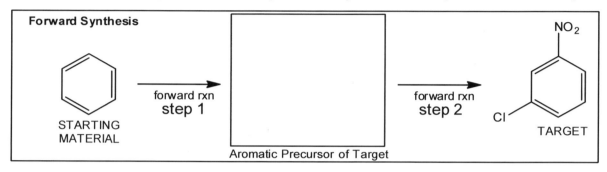

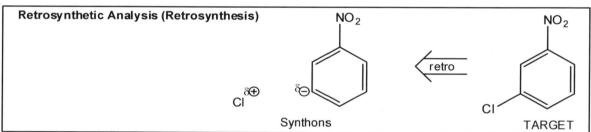

Construct Your Understanding Questions (to do in class)

1. This synthesis is simple enough solve in the forward direction by writing the reagents for each step over the reaction arrows, but for practice let's think backwards from the TARGET.

    a. In the "Retrosynthetic Analysis" section at the bottom of Model 1, put an **X** through the bond on the target that would give the **synthons** shown.

    b. The **precursors** associated with these synthons are <u>stable organic reagents that behave like the synthons</u>. In the box above the synthons, draw the precursor of the aromatic synthon shown.

    c. (Check your work.) In an EAS reaction involving nitrobenzene, a carbon *meta* to the NO_2 group will function as a nucleophile. Is this consistent with your answer to part b?

    d. Over the reaction arrow labeled Step 2, draw a precursor of the synthon Cl^+ (that is, reagent(s) that will act like Cl^+ and can be mixed with nitrobenzene to give the target, *meta*-chloronitrobenzene).

2. Over the <u>reaction arrow</u> labeled Step 1, write reagents that will transform benzene into nitrobenzene, and over the Step 2 arrow write reagents that will give the Target.

3. In our retrosynthetic analysis in Model 1, we guessed that the C—Cl bond would be formed from an aromatic carbon nucleophile and an electrophilic chlorine. Draw synthons with the opposite polarity (i.e., $Cl^{\delta-}$ and $C^{\delta+}$). Which strategy seems better?

4. Is your answer to the previous question consistent with the fact that we have not learned any reaction in which a nucleophilic chlorine atom makes a bond to an (electrophilic) aromatic carbon atom?

5. Now consider what happens if you start this retrosynthetic analysis by putting an **X** on the C—N bond (instead of the C—Cl bond). Explain why the second step in the resulting synthesis fails even though there are reagents (H_2SO_4/HNO_3) for adding a nitro group to an aromatic ring.

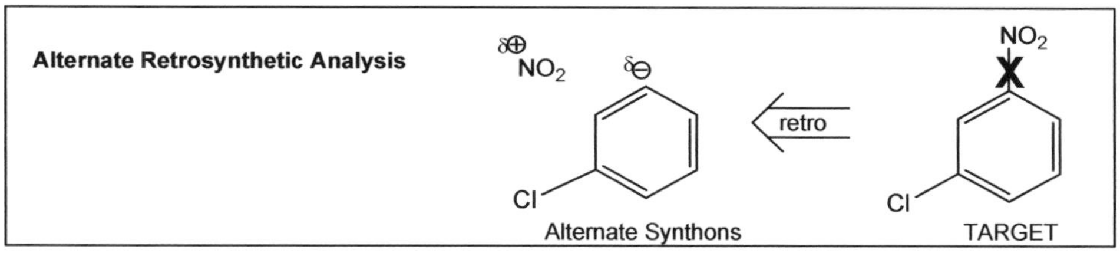

Alternate Retrosynthetic Analysis

Alternate Synthons

TARGET

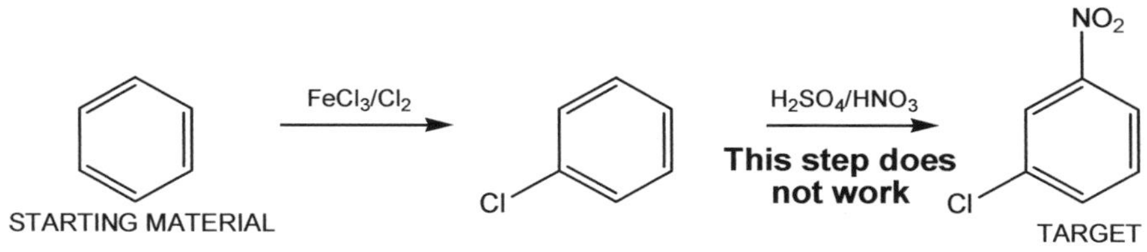

STARTING MATERIAL

FeCl₃/Cl₂

H₂SO₄/HNO₃

This step does not work

TARGET

6. List problems you might encounter in a synthesis of the following target from benzene.

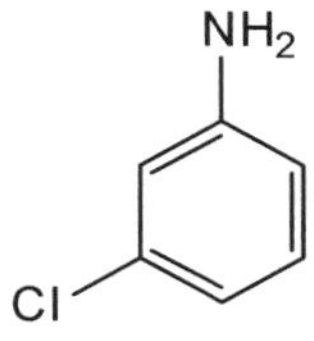

Model 2: Changing an *o/p* to a *m* Director, or Vice Versa

It is sometimes possible to change an *ortho/para* director into a *meta* director, or vice versa.

Synthetic Transformations 3.1, 5.1, & 5.2: EAS Directing Group Changes

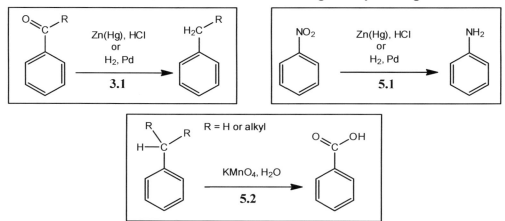

Construct Your Understanding Questions (to do in class)

7. Label the group on each of the six benzene derivatives in Model 2 as *o/p* **director** or *m* **director**.

8. Label each reaction in Model 2 as an **oxidation** or a **reduction** (of the aromatic molecule).

9. In retrosynthesis, a retro arrow can be associated with crossing out a bond (as you did in Model 1) or with a functional group transformation as shown below. Add a reaction arrow (from left to right) to the step below, and write in reagents that will transform the precursor into the target.

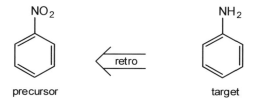

precursor   target

10. Use retrosynthesis to design a synthesis of the following target from benzene as the starting material. Be sure to write reagents over the reaction arrow for each forward step.

11. Look back at your answer to Question 6 and explain the utility of the Synthetic Transformations in Model 2.

Model 3: Attenuating the Effects of a Strong Activator *(attenuate = weaken)*

It turns out that aniline is a problematic starting material for an EAS reaction. The NH_2 group is such a strong activator that many EAS reactions will lead to multiple substitutions, as shown at right.

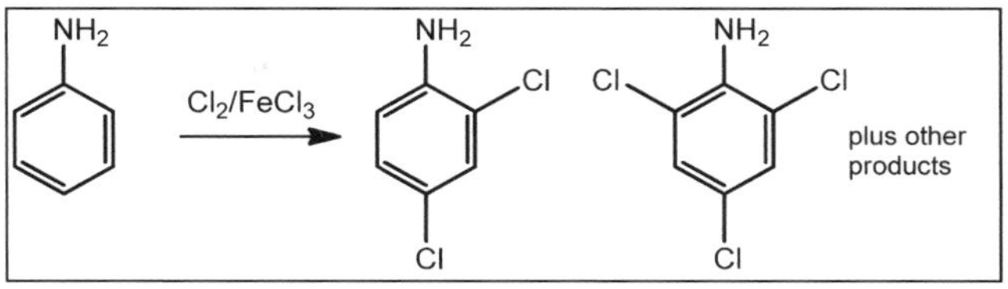

Other EAS reactions (such as Friedel-Crafts reactions) do not work at all because the NH_2 group is a decent nucleophile and can undergo an S_N2 reaction instead.

Synthetic Transformations 5.3 and 5.4: Acylation and de-Acylation of Anilines

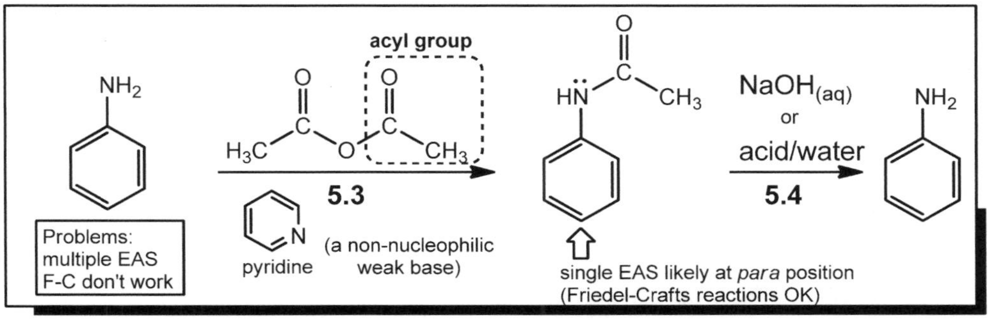

The mechanisms of both these reactions will be covered in a later ChemActivity.

Extend Your Understanding Questions (to do in or out of class)

12. Circle the **acyl group** on the acylated aromatic amine in Model 3.

13. Draw a second-order resonance structure demonstrating that the lone pair shown on the N of the acylated aniline is pulled (delocalized) toward the oxygen.

14. Construct an explanation for why, unlike aniline, an acylated aniline can undergo a single controlled EAS reaction, and explain why substitution is favored at the *para* position over the *ortho* position.

15. Design a synthesis of the following target from benzene.

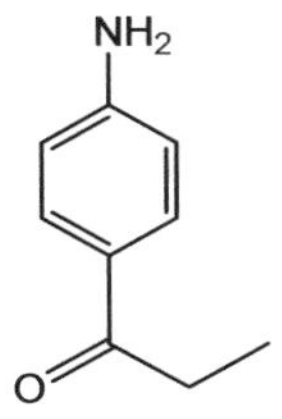

Model 4: Protecting Groups

A **sulfonic acid group** can be added or removed from an aromatic ring. This means an SO_3H group can be used to temporarily block a position on an aromatic ring during a synthesis.

Synthetic Transformations 2.3 & 5.5: Sulfonation and de-Sulfonation of Aromatic Rings

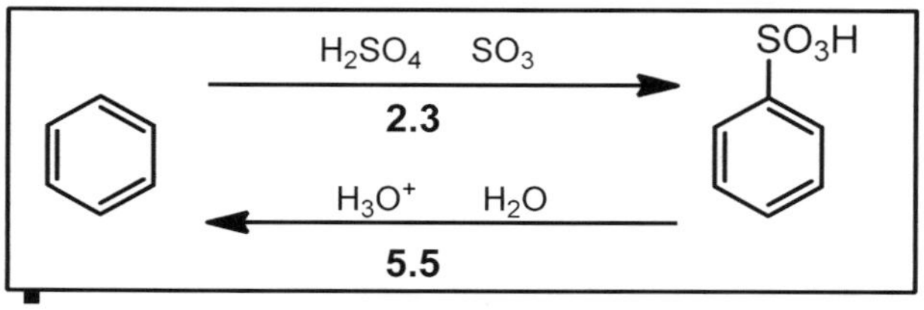

Extend Your Understanding Questions (to do in or out of class)

16. In EAS involving a bulky substituent (bulky is usually defined as $2°$ or $3°$) the electrophile is directed by steric effects to the *para* position over the *ortho* position (as shown below).

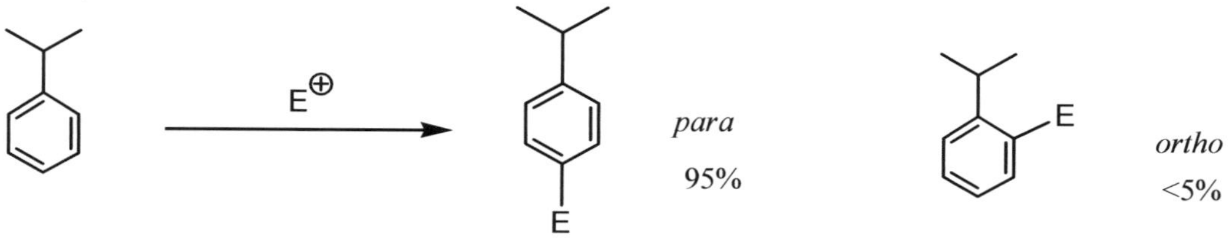

Propose a synthesis from this starting material that gives **primarily the *ortho*-nitro product** shown below. *Hint*: Re-read the underlined statement in Model 4.

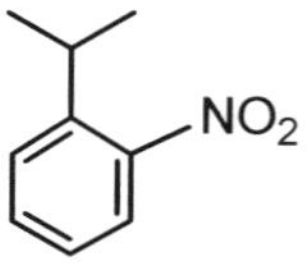

TARGET

17. Construct an explanation for why the sulfonic acid group in your synthesis in the previous question is said to be used as a "**protecting group.**"

Model 5: Nucleophilic Aromatic Substitution

In your course you will likely encounter a reaction that is very different from EAS, but has a name that sounds similar: Nucleophilic Aromatic Substitution (NAS) is in many ways the opposite of EAS, and the two reactions are frequently confused with one another. The mechanism is not covered here, but it may help to know the following about NAS reactions...

Memorization Task 5.1: For NAS to be favorable...

- There must be a leaving group on the aromatic ring (usually a halogen).

- You must use a very good nucleophile such as RO^- or R_3N (R = H or alkyl).

- There must be at least one strong **electron-withdrawing group (EWG)** *ortho* or *para* to the leaving group (usually a resonance EWG like nitro or a carbonyl).

- The more EWG's *ortho* and/or *para* to the leaving group the faster the rate of NAS.

*Special NAS reactions can take place without electron-withdrawing groups on the ring. These require extreme conditions and likely go through what is called a **benzyne intermediate**.*

Synthetic Transformation 5.6: NAS (Nucelophilic Aromatic Substitution)

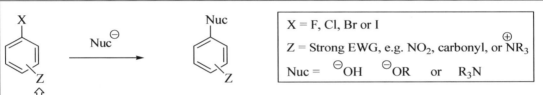

X = F, Cl, Br or I

Z = Strong EWG, e.g. NO_2, carbonyl, or $\overset{\oplus}{NR_3}$

Nuc = $^{\ominus}OH$ $^{\ominus}OR$ or R_3N

Under normal conditions there must be at least one EWG *ortho* or *para* to the leaving group. (Certain conditions do not require an EWG. See Synth. Transf's 21.2 and 21.3 in Exercises)

Extend Your Understanding Questions (to do in or out of class)

18. Match each term to a molecule to indicate whether it will undergo NAS, EAS, or neither.

moderate NAS	moderate EAS	neither (*use twice*)
fast NAS	fast EAS	

19. Make a list of key similarities and differences between NAS and EAS reactions.

Confirm Your Understanding Questions (to do at home)

20. *tert*-Butyl benzene does not react with $KMnO_4$. Is this consistent with Synthetic Transformation 5.2? Explain.

21. Predict the product of each of the following reactions.

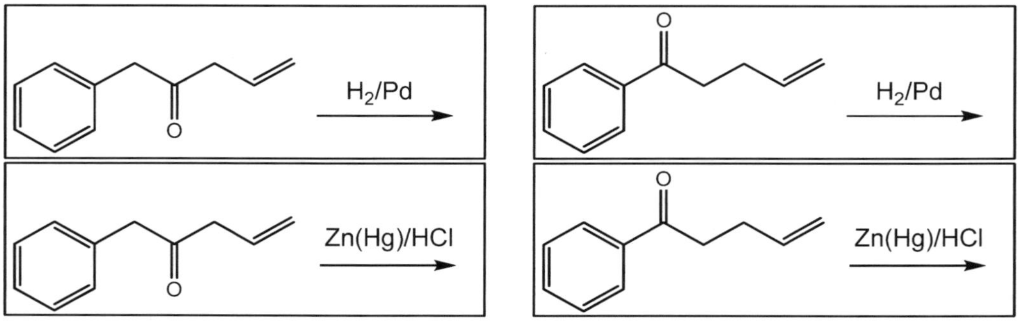

22. Design a synthesis of *n*-propylbenzene from benzene. *Hint*: A one-step synthesis using $CH_3CH_2CH_2Br$ and $AlBr_3$ yields mostly isopropylbenzene.

23. Design a high-yield synthesis of each target starting from benzene.

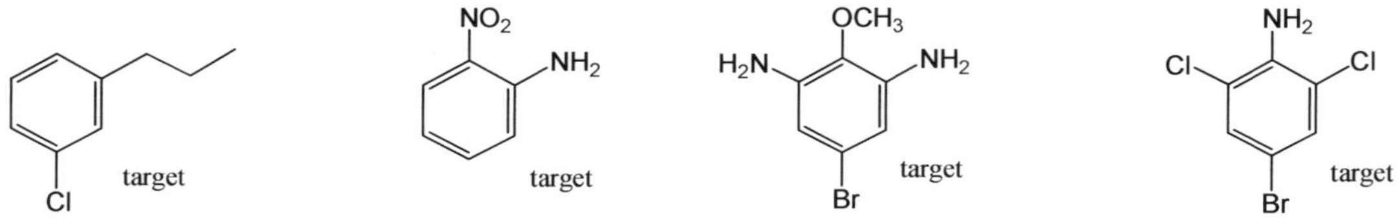

24. Make up a synthesis problem that focuses on synthetic transformations involving aromatic molecules. If it is a good synthesis problem, email it to your instructor (write a description of it in words), and there is a chance it will show up on the next quiz.

25. Construct an explanation for why neither of the aromatic rings at right will undergo NAS even though each has two resonance electron-withdrawing groups.

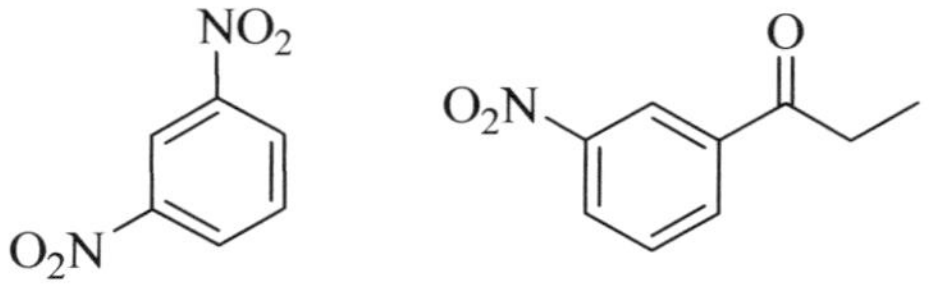

Read the assigned pages in your text, and do the assigned problems.

The Big Picture

Synthesis problems involving aromatic rings usually include EAS reactions. The new reactions in this activity are not EAS reactions, but they can be very useful for directing substituents in an EAS reaction or modifying substituents after they have been added to a ring in an EAS reaction.

This activity also introduces the concept of a temporary group that is used to reduce (attenuate) the reactivity of an aryl amine or block a position on the ring. These are generally called **protecting groups**. Protecting groups play a huge role in syntheses performed in the laboratory.

This activity is an opportunity to sort out the differences between NAS and EAS. Both are useful in the synthesis of aromatic molecules. The next group of ChemActivities will focus on the chemistry of carbonyl compounds.

Common Points of Confusion

Many students consider synthesis to be the most challenging (and frustrating) part of an organic course. The following advice is offered on how to make synthesis less intimidating (and perhaps even fun):

In organic synthesis, as with every topic in this course, the situation is not nearly as bad as you think. Do not let the reputation of this course or your anxiety get in the way of enjoying organic synthesis, one of the most creative and potentially fun challenges you will encounter in all of science.

- **Memorize all the Synthetic Transformations**. Many students find making notecards to be useful.. You must know these forward and backward. Failure to adequately memorize the Synthetic Transformations is the most common cause of difficulties with synthesis problems.

- Practice doing synthesis problems working both from the starting material forward and, using retrosynthesis, from the target backwards.

- Don't waste time staring at a blank page. Keep your pencil moving and your brain will follow. Start each synthesis by listing the possible first steps forward from the starting material AND the possible last steps, thinking backwards (retrosynthetically) from the target.

- Many two- to four-step synthesis problems can be solved in the forward direction without using retrosynthesis, but retrosynthesis is an invaluable tool for developing a plan for solving complicated syntheses. If you do not understand the point or methods of retrosynthesis, discuss it with a person who seems to like synthesis.

- Chess masters study games played by other masters to learn common sequences of moves. Study the solutions to synthesis problems to build your repertoire of common sequences of transformations.

- Make up your own synthesis problems, and share them with your study partner.

Notes

ChemActivity 6: Organometallic Reagents

(What are the special conditions required for preparation and use of Li and Mg organometallics?)

Model 1: Lithium Reagents and Grignard Reagents

We have learned so far in this course that R_3C^- is very high in potential energy. It turns out that carbon does not form useful salts with the usual **counterions** (e.g., $NaCH_3$ and KCH_3 have no synthetic utility).

However, it does form useful salt-like reagents with some metals, including lithium and magnesium. These are examples of **organometallic reagents**, so called because they have a metal-carbon bond.

Lithium and Grignard reagents are prepared from alkyl halides by a complex mechanism. Though there are differences between them, we <u>assume in this activity that lithium and Grignard reagents are interchangeable</u>, each behaving like a carbon anion, that is, an **excellent nucleophile** and an **extremely strong base** (releasing about **35 pK_a** units of energy upon forming a new bond to an acidic H).

"Grignard Reagents" are named for French chemist Victor Grignard (1871-1935), pronounced "green-yard."

Synthetic Transformation 6.1: Lithium or Grignard reagent from an alkyl halide

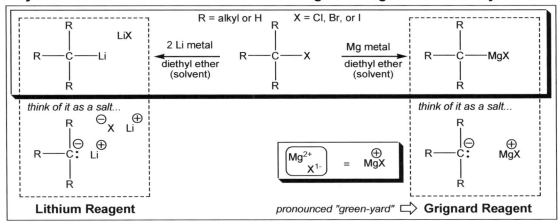

Lithium Reagent   pronounced "green-yard" ⇨ **Grignard Reagent**

Construct Your Understanding Questions (to do in class)

1. Consider the lithium and Grignard reagents in Model 1.

   a. Given its position on the periodic table, what is the likely positive charge on a Li ion?

   b. Given its position on the periodic table, what is the likely positive charge on a Mg ion?

   c. What is the charge on a Br ion?

   d. What is the total charge on a combined [MgBr] ion?

   e. Is your answer to the previous question consistent with the fact that a Li or [MgBr] ion can act like a counterion for a carbon with a -1 charge?

Memorization Task 6.1: Think of lithium and Grignard reagents as salts.

You will see these reagents drawn as having a covalent bond between C and Li or MgBr (as in the top representations in Model 1), but they behave as if there is a −1 charge on carbon and a +1 charge on Li or MgBr. It may help to draw the reagents as salts (as in the bottom representations in Model 1).

Synthetic Transformation 6.2: Reaction of a lithium or Grignard reagent as a base

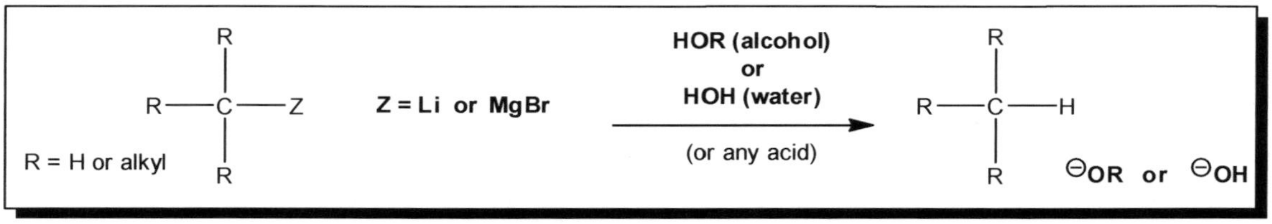

R—C(R)(R)—Z   Z = Li or MgBr   → HOR (alcohol) or HOH (water) (or any acid) →   R—C(R)(R)—H $^{\ominus}$OR or $^{\ominus}$OH

R = H or alkyl

Memorization Task 6.2: Use only <u>unfunctionalized</u> R groups to make R—Li or R—MgX.

Because of their extreme basicity, lithium and Grignard reagents CANNOT be made from an alkyl halide (R—X) containing an H with a pK_a less than 35. For the same reason, water interferes with the preparation and use of these reagents. Flame-dried glassware is often used for these reactions.

2. Circle the two alkyl halides below that are suitable for making a lithium or Grignard reagent.

H_3C—C≡C—CH_2Br

(cyclopentane with Cl and alkyne substituents)

HO————I

(benzene ring with Br)

H_2N————Br

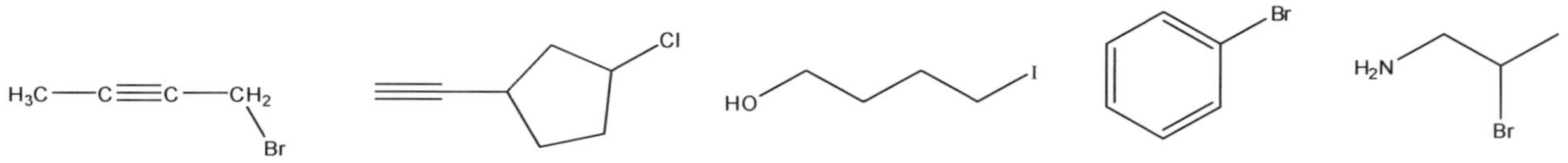

3. Draw the Grignard reagent formed by mixing bromocyclohexane with magnesium metal. (As in Model 1, show it two different ways, with a covalent bond between C and MgBr and as a salt.)

a. Use curved arrows to show an acid-base reaction between water and **each representation** of the Grignard you drew above.

b. List at least one reagent could you mix with the Grignard above to produce monodeuterocyclohexane? → →

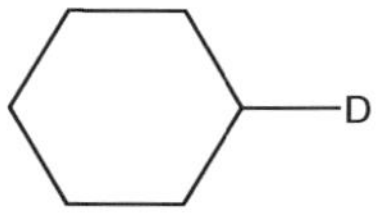

(cyclohexane with —D)

Synthetic Transformation 6.3: Replacement of a halogen with a deuterium

$$
R{-}\underset{\underset{R}{|}}{\overset{\overset{R}{|}}{C}}{-}X \xrightarrow[\substack{\text{diethyl ether}\\\text{(solvent)}}]{\text{Li or Mg metal}} \xrightarrow{\text{D}_2\text{O}} R{-}\underset{\underset{R}{|}}{\overset{\overset{R}{|}}{C}}{-}D \quad {}^{\ominus}\text{OD}
$$

R = H or alkyl

Model 2: Carbonyl (C=O) Compounds as Carbon Electrophiles

A functional group containing a **carbonyl group** (C=O) is called a **carbonyl compound**. So far we have encountered aldehydes, ketones, and carboxylic acids. Later we will encounter several other common variations.

Due to the polarization of the C=O bond, the C of a carbonyl group has a $C^{\delta+}$, and often functions as an electrophile. Note: There are two main types of carbon electrophiles, **carbonyl compounds** ($Z_2C{=}O$) and **alkyl halides** ($R_3C{-}X$). Both serve as the synthetic equivalent for a synthon with a $C^{\delta+}$.

Though lithium and Grignard reagents are good nucleophiles, for reasons we will not discuss they do not give good product yields in reactions with **alkyl halide electrophiles** (as shown below), but…

low yield of this product

…they do react with carbonyl compounds (as shown in Memorization Task 6.3).

Memorization Task 6.3: Lithium and Grignard react well with C=O electrophiles

---

Construct Your Understanding Questions (to do in class)

4. The carbonyl compound in Memorization Task 6.3 is an example of an **aldehyde**. Add a δ+ and a δ− to this molecule to show the polarization of the carbonyl (C=O) bond.

5. In the box above, draw the product that results from the curved arrows shown. (A "bouncing" curved arrows is used here to emphasize that the bond made from the electrons in the C—MgBr bond goes from the C (not the MgBr).

6. Model 2 shows the carbon-metal bond of the organometallic reagent (a Grignard in that case) as a covalent bond, but it can be more illustrative to show it an ionic bond. Use curved arrows to show how the organometallic reagent CH_3—Li (shown below as a salt) might react with a carbonyl compound, and draw the resulting products.

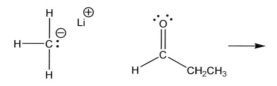

Model 3: Nucleophilic Addition to a Carbonyl (C=O)

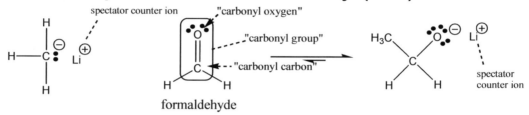

formaldehyde

Note: The product is sometimes shown with a covalent bond between O and Li.

Construct Your Understanding Questions (to do in class)

7. Add curved arrows to Model 3 that account for the product shown, then check your work in the previous question.

8. The **alkoxide** (RO⁻) product in Model 3 is itself a strong base and a good nucleophile.

   a. Draw the end product that results if the product shown in Model 3 is neutralized with dilute acid (H_3O^+/H_2O).

   b. Draw the product that results if the product shown in Model 3 is treated with the alkyl halide CH_3CH_2—I (instead of dilute acid).

Synthetic Transformations 6.4 and 6.5: Alcohol or ether from an aldehyde or ketone

$$R_3C{-}Z$$
$$Z = Li \text{ or } MgX$$
diethyl ether (solvent)

$$H_3O^+/H_2O \text{ (dilute acid)}$$

$$R = H \text{ or alkyl}$$

$$R_3C{-}Z$$
$$Z = Li \text{ or } MgX$$
diethyl ether (solvent)

$$R''{-}\overset{H_2}{C}{-}X$$
(1° alkyl halide)

$$R = H \text{ or alkyl}$$

Extend Your Understanding Questions (to do in or out of class)

9. Draw the most likely product of the following sequence of reactions starting with cyclohexane.

cyclohexane $\dfrac{Cl_2 \;\; h\nu}{\text{one molar equiv. Cl}}$ $\dfrac{Li^0}{\text{diethyl ether (solvent)}}$ $O{=}C(CH_3)_2$ (acetone) $CH_3CH_2CH_2Br$

Model 7: Reduction of Aldehydes or Ketones to Alcohols

Many nucleophiles will react with a carbonyl carbon. We will explore more possibilities in future activities. Here we discuss one small but very important nucleophile, **hydride anion (H⁻)**.

As with carbon, the sodium and potassium salts of hydrogen (NaH and KH) are not nearly as useful as other reagents that deliver an H^-. The most common are $LiAlH_4$ and $NaBH_4$

Memorization Task 6.3: LiAlH₄ and NaBH₄ react as if they were a hydride anion (H⁻)

LiAlH₄ (lithium aluminum hydride) = strong H^- delivery reagent (reacts with any carbonyl)

NaBH₄ (sodium borohydride) = milder H^- delivery reagent (reacts with only aldehydes and ketones)

Extend Your Understanding Questions (to do in or out of class)

10. Use curved arrows to show the mechanism, and draw the products of the most likely reaction between a hydride anion (H⁻) and the ketone below.

$$\overset{\ominus}{H}:$$

11. (Check your work.) The product of a reaction in the previous question should be an alkoxide (RO⁻), a strong base and a good nucleophile.

   a. Draw the result if the product from the reaction in the previous question is treated with dilute acid (H_3O^+/H_2O).

   b. Is your answer to a consistent with Synth. Transf. 6.6?

Synthetic Transformation 6.6: Reduction of an aldehyde or ketone to an alcohol

   c. Draw the product that results if the product from the reaction in the previous question is treated with the alkyl halide $CH_3CH_2CH_2$—I (instead of dilute acid).

Confirm Your Understanding Questions (to do at home)

12. Draw a precursor of ....

    a.   an aldehyde (RHC=O)

    b.   a ketone (R_2C=O)

    c.   a carboxylic acid (RCO_2H)

    d.   a bromohydrin (Br and OH on adjacent carbons)

13. Show a synthesis of 2-deutero-2-methylpropane from 2-methylpropane.

14. Use retrosynthesis to design a synthesis of the following molecule from bromobenzene.

15. Lithium and Grignard reagents are extremely strong bases. Upon making a bond to H^+ they release a large amount of energy (you may assume about 50 pKa units, but actually the value is closer to 35–40 pKa units of energy). Recall that the pKa of an alcohol is about 6. (Requires 16 pKa units of energy to remove an H^+.)

    a.   In the reaction below, the moment a molecule of Grignard reagent is formed, it immediately undergoes an acid-base reaction with a second molecule of starting material. Use curved arrows to show this unwanted side reaction, and draw the products.

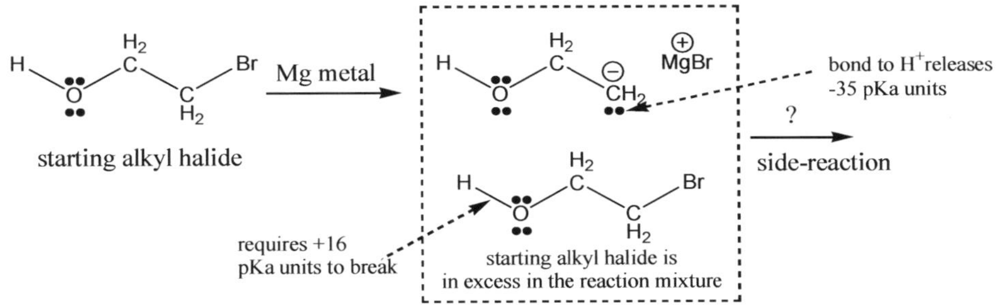

    b.   Construct an explanation for why preparation of lithium and Grignard reagents **cannot** be performed with an alkyl halide containing an H with a pK_a < 35.

16. How does the mechanism of nucleophilic addition differ from nucleophilic substitution? Do the names reflect these differences?

17. Which of the following are unacceptable alkyl halides for making a Grignard or lithium reagent?

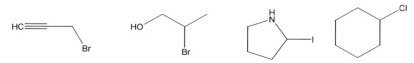

Rule of Thumb: In general, an alkyl halide with any other functional group is not suitable for making a Grignard or lithium reagent.

18. For each unacceptable alkyl halide in the previous question, show the side reaction that would occur if each were treated with Li or Mg.

19. Draw the alcohol product that results when an aldehyde or ketone is…

    a.   treated with a Grignard or Lithium reagent (strong nucleophile)

    b.   then neutralized with dilute acid

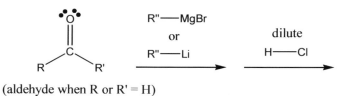

    (aldehyde when R or R' = H)

20. Design a synthesis of each target using the starting material and any reagents containing three carbons or less.

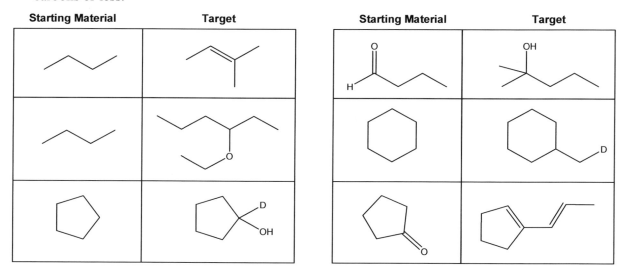

Read the assigned pages in the text, and do the assigned problems.

The Big Picture

One of the biggest challenges of synthetic organic chemistry is the formation of carbon-carbon bonds to form the correct carbon backbone of the molecule. We will only learn a few ways to make carbon-carbon bonds in this course.

Research in synthetic organic chemistry is largely devoted to the discovery of new and exotic ways to make carbon-carbon bonds. If you take an advanced synthetic organic chemistry course you will learn many of these methods.

This year's Nobel Prize in Chemistry went to three pioneers of the use of the metal palladium to form new carbon-carbon bonds: Richard F. Heck of Delaware University, Ei-ichi Negishi of Purdue University, and Akira Suzuki of Hokkaido University (Japan).

Common Points of Confusion

- Grignard and lithium reagents are often represented as covalent molecules. This can be confusing. A good strategy (at least at the start) is to draw these organometallic reagents as salts. (A "salt" is simply an ionic compound consisting of a + and a − ion.) By drawing the Grignard or lithium reagent with a − charge on carbon, you emphasize the nucleophilic and basic nature of this carbon.

- A persistent error among organic chemistry students is the belief that an aldehyde or ketone (C=O) can be transformed into an alcohol via electrophilic addition [using acid/water]. This is not the case. A reducing agent such as $LiAlH_4$ or $NaBH_4$ is required.

- Students also assume that the opposite reaction, transforming an alcohol (C—OH) into an aldehyde or ketone (C=O) can be accomplished via elimination [using a strong base, for example]. This is not the case. An oxidizing agent such as chromic acid must be used.

Notes

ChemActivity 7: Addition-Elimination

(What is the mechanism of imine, enamine and acetal formation?)

Model 1: Amine Addition-Water Elimination (Imines and Enamines)

Synthetic Transformation 7.1: Imine Formation

aldehyde or ketone   →   imine

Synthetic Transformation 7.2: Enamine Formation

aldehyde or ketone   →   enamine

Construct Your Understanding Questions (to do in class)

1. Devise a five-to-six step mechanism for Synthetic Transformation 7.1. *Hint*: Recall that the carbon of C=O is electrophilic (δ+ charged) an OH group can be protonated with acid to make it into a good leaving group (water).

2. (Check your work.) <u>Either</u> of the following is a logical first step for the reactions in Model 1, though the "Alternate Step 1" does not happen for reasons discussed in Common Points of Confusion section. Check your first step above.

3. What is the difference between **imine**-forming reagents and **enamine**-forming reagents?

4. The following is an incomplete mechanism for Synthetic Transformation 7.2 (enamine formation).

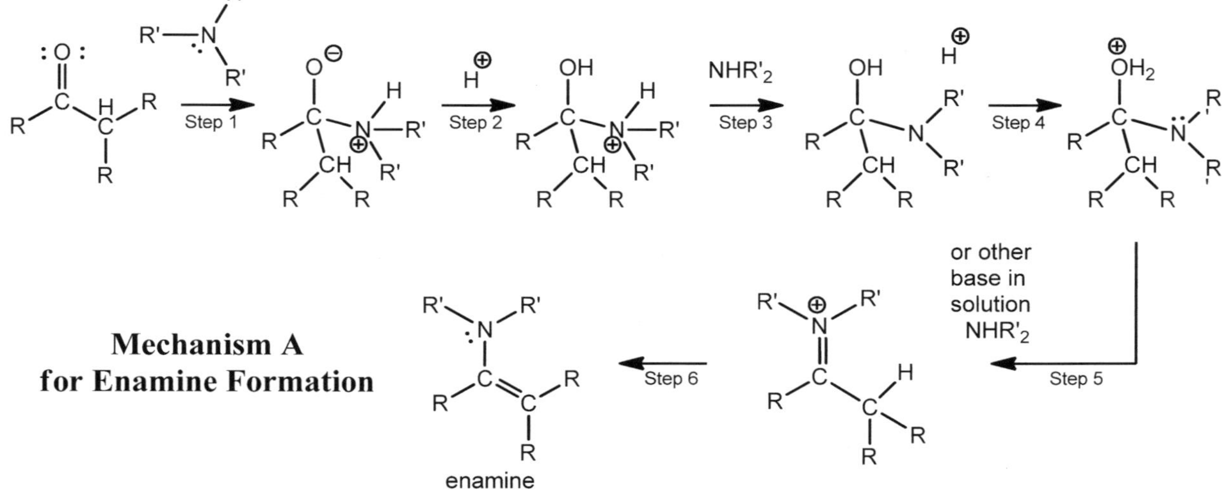

**Mechanism A
for Enamine Formation**

enamine

a. Complete the mechanism above by adding curved arrows. (All intermediates are shown.)

b. (Check your work.) Except for the last step, enamine formation is identical to imine formation. Note any differences between this and your imine mechanism (previous page).

5. Mechanism B (below) is yet another alternate acceptable mechanism for enamine formation.

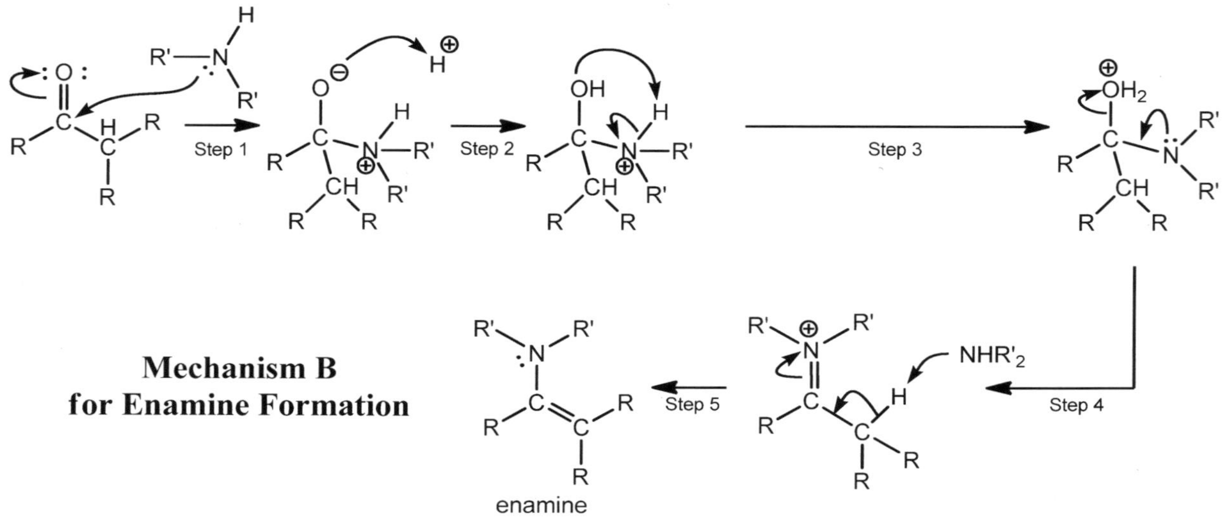

**Mechanism B
for Enamine Formation**

enamine

a. Where does Mechanism B differ from Mechanism A?

b. Construct an explanation for why Step 3 is called an **intramolecular proton transfer**.

c. Typically, intramolecular proton transfers like Step 3 are unlikely because they require a highly strained four-member ring transition state. Draw this strained transition state.

Model 2: Alcohol Addition-Water Elimination (Hemiacetals & Acetals)

Synthetic Transformation 7.3: Hemiacetal and Acetal Formation

Critical Thinking Questions

6. [E]Acetal formation is neither uphill nor downhill in terms of energy. According to Model 2, what are two ways to drive this reaction toward acetal formation (even though it is not downhill)?

7. Explain why both of these actions promote acetal formation.

8. Show the mechanism of the first half of Model 2 (hemiacetal formation). Hint: the first step involves the acid catalyst. (On the next page you are asked to generate a mechanism for the second step.)

9. (Check your work.) Hemiacetal formation is nearly identical to hydrate formation (below). A hydrate is what forms if you add large amounts of water to an aldehyde except ROH is used in place of HOH. Check your mechanism above against the one below.

Extend Your Understanding Questions (to do in or out of class)

10. Show the mechanism of the second step in Model 2, acetal formation (from a hemiacetal).

11. Draw the end product(s) of each reaction. For the first seven reactions with an* *assume water is distilled (removed) from the product mixture during the reaction*.

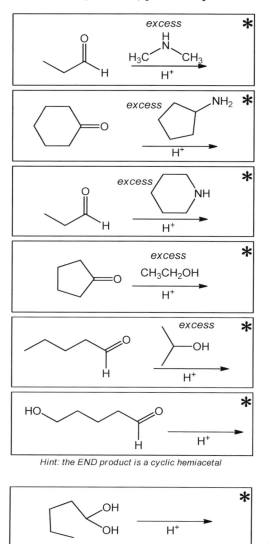

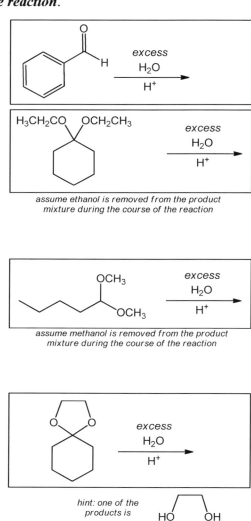

Model 3: Cyclic Acetal Protecting Groups

Synthetic Transformation 7.4: Protection of a Carbonyl Against Nucleophiles

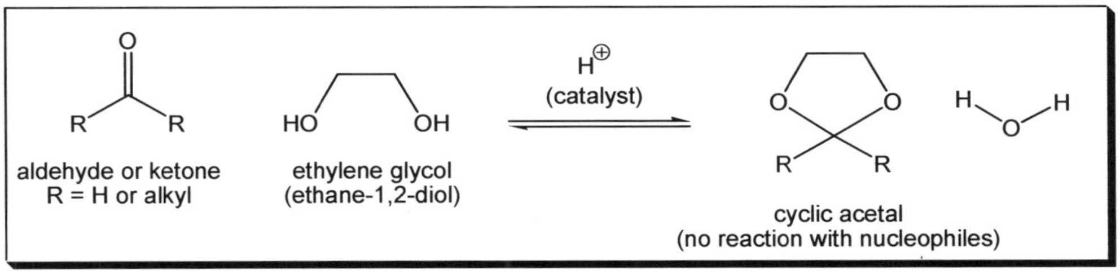

The mechanism of this reaction is explored in the homework for this ChemActivity

- Cyclic acetals are very stable in the presence of most bases and nucleophiles (without acid).

- Treatment of a cyclic acetal with acid and excess water generates the original aldehyde or ketone.

Extend Your Understanding Questions (to do in or out of class)

12. The reaction in Model 3 is neither uphill nor downhill. How can you drive this reaction…

    a. to the right, producing a high yield of cyclic acetal?

    b. to the left, producing a high yield of the aldehyde or ketone starting material?

13. [E]According to Model 3, which is more susceptible to reaction with a nucleophile (e.g., RNH$_2$, H$_3$C-Li, RC≡CNa, KC≡N, etc.), **an aldehyde/ketone** or a **cyclic acetal** (circle one)?

14. Construct an explanation for why the following reaction will NOT give a significant yield of the target shown. Then draw the nucleophilic addition product that will form.

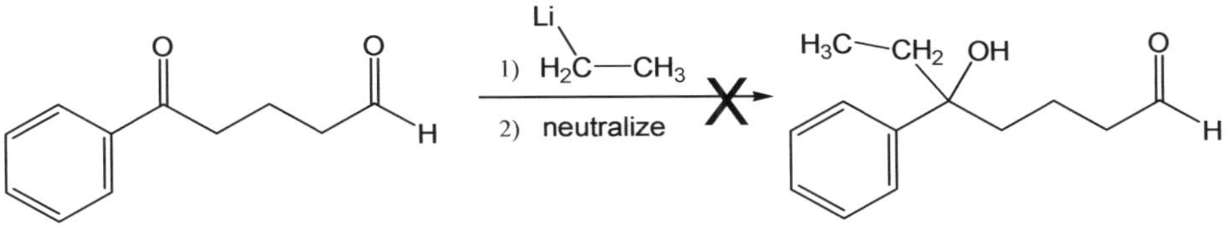

15. (Check your work) Is your answer to the previous question consistent with the following data? Explain.

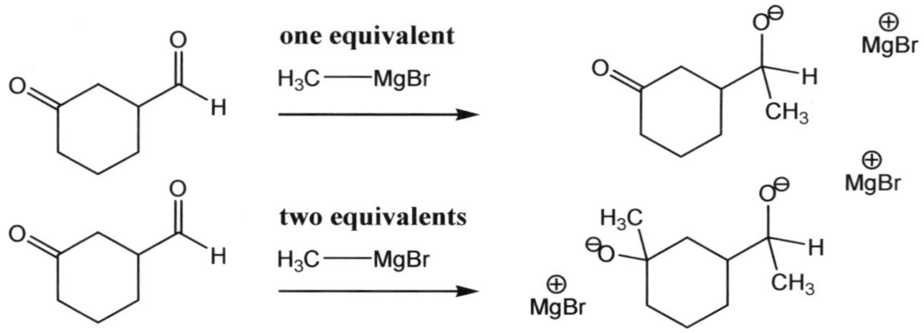

16. Based on the reactions in Model 10, which is **more reactive** toward nucleophilic addition, an **aldehyde carbonyl** or a **ketone carbonyl** (circle one)?

17. Explain why the product in the box forms almost exclusively under these conditions.

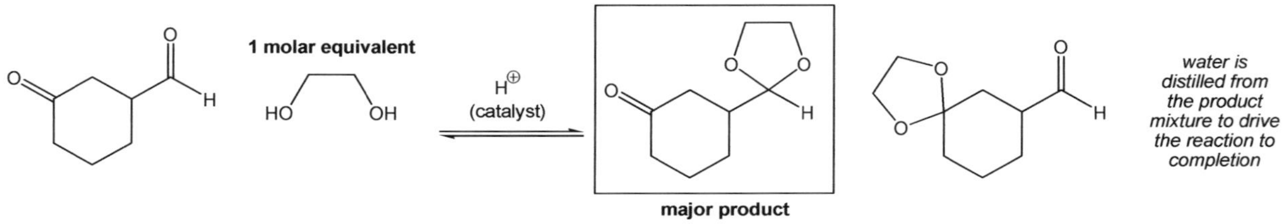

major product

water is distilled from the product mixture to drive the reaction to completion

18. Design a way to accomplish the following synthesis (from the previous page). *Hint*: The structure in the box above is formed during this synthesis.

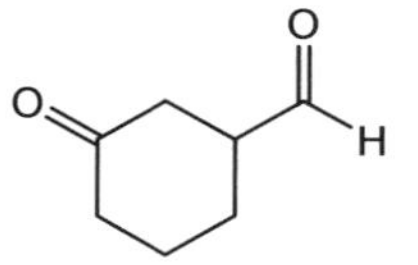

starting material

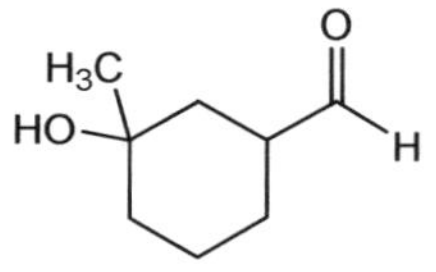

target

19. (Check your work) Construct an explanation for why the molecule ethan-1,2-diol is considered a **protecting group** in your synthesis above.

Confirm Your Understanding Questions (to do at home)

20. Construct a mechanism for the following reaction.

butanal   (catalyst)

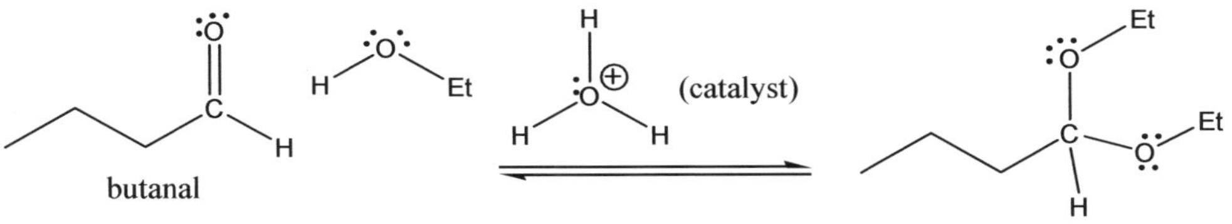

a. Identify the hemiacetal that forms during the reaction above.

b. What could you do to push the above reaction toward a high yield of this acetal?

21. Show the mechanism by which the following acetal can be converted into an aldehyde, and draw the resulting aldehyde.

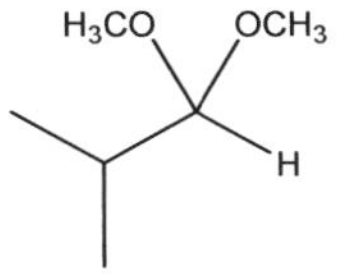

22. Note that some textbooks use the term "acetal" to describe only the product that forms when an alcohol and an acid catalyst are mixed with an aldehyde, and the term "ketal" to describe the product that forms when an alcohol and an acid catalyst are mixed with a ketone. Based on this terminology, draw an example of each of the following: hemiacetal, acetal, hemiketal, ketal.

23. Design a mechanism for each of the following reactions

a)

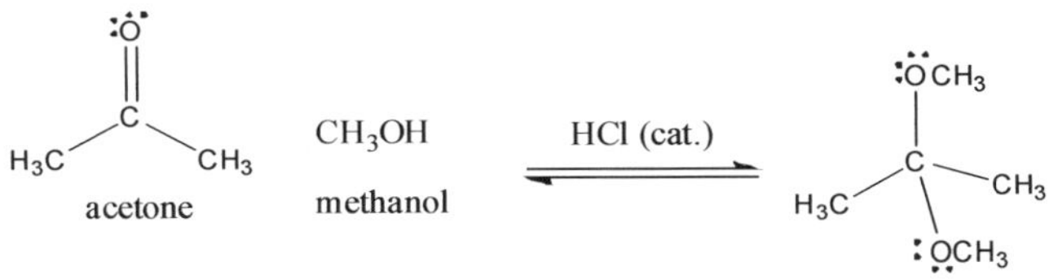

acetone   methanol

b)

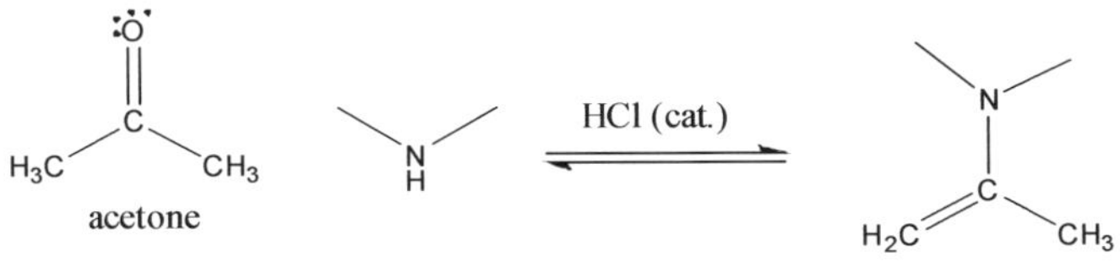

acetone

c)

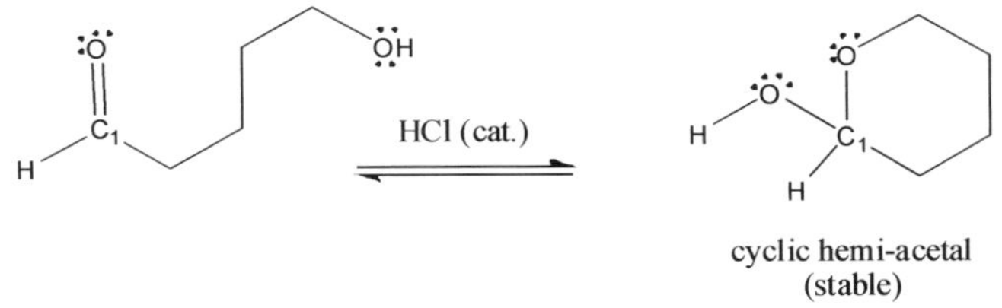

cyclic hemi-acetal
(stable)

d)

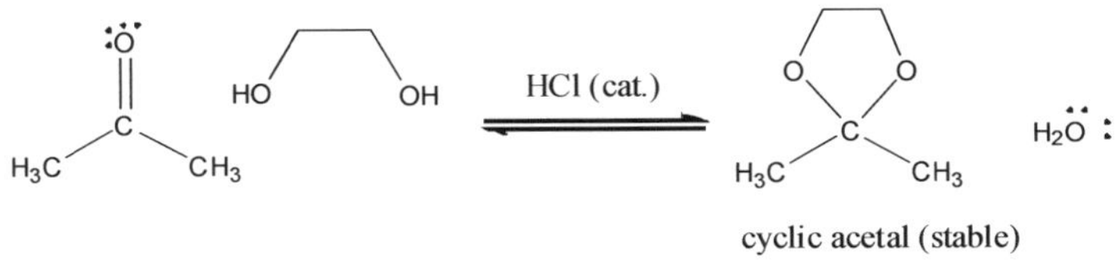

cyclic acetal (stable)

24. Construct a mechanism for each of the following transformations.

25. What reagents could you add to the imine above to turn it back into the corresponding ketone?

26. The optimal pH for an imine or enamine reaction is 4.5. Construct an explanation for why these reactions do not work at very low pH or at very high pH.

27. When both are possible, an imine formation (C=N bond forms in the last step) is favored over an enamine formation (C=C bond forms in the last step). Explain why reaction with a secondary amine cannot lead to formation of a C=N bond (an imine) in the last step.

28. Design a synthesis of each target from the starting material given.

29. Design a synthesis of the target from the starting material shown.

30. The following is an alternate use of a cyclic acetal as a protecting group.

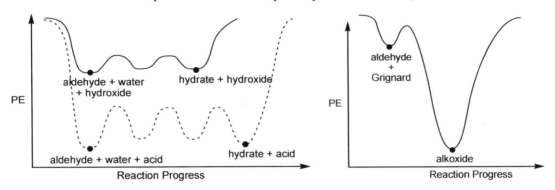

a.   Identify the cyclic acetal protecting group, and circle the functional group(s) in the starting material that the cyclic acetal protects.

b.   From what reagent does the cyclic acetal protect these functional groups?

c.   What side product(s) might form if the starting material were treated directly with NaH, then CH₃I (without using the protecting-group strategy outlined above)?

d.   Devise a mechanism for the first step in this synthesis.

31. Devise a mechanism for base-catalyzed hydrate formation.

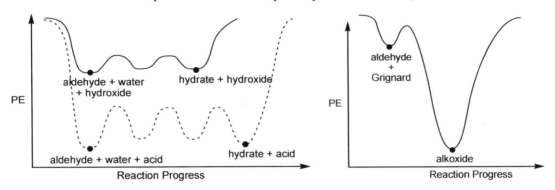

32. Consider the following reaction profile for base-catalyzed hydrate formation (drawn on the same scale as the reaction profile for acid-catalyzed hydrate formation).

a.   (Check your work.) Does the reaction profile fit your mechanism? Explain your reasoning.

b.   Compare the two reaction profiles on the left to the one on the right for reaction of an aldehyde with a Grignard, and construct an explanation for why only the reaction with the Grignard is steeply downhill (exothermic).

c.   Construct an explanation for why the two energy profiles on the left are considered "reversible" while the one on the right is considered "irreversible."

33. The reagent sodium borohydride (NaBH₄) reacts exactly the way you would expect H⁻ to react (H⁻ is a very strong base and is high in energy like a Grignard or lithium reagent). Given this, predict the product that results when butanal is treated with NaBH₄ followed by dilute HCl.

34. What reagents could you use to convert this alcohol back to butanal? *Hint*: The sodium borohydride reaction in the previous Exercise is NOT reversible and is considered a reduction of butanal.

35. Construct a mechanism for the following reaction.

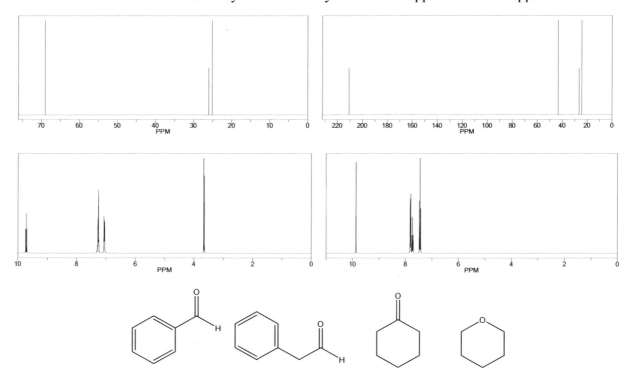

36. Review and memorize the following facts about NMR and of aldehydes and ketones, and use this information to match each spectrum to the correct structure. (Each spectrum is of a different compound).

   a. In H^1 NMR, an aldehyde H shows up near 10ppm (to the far left, also called "downfield").

   b. In C^{13} NMR the carbonyl C of an aldehyde or ketone appears above 200ppm.

37. Explain why the following simple synthesis of the target shown will not work.

38. Consider the following synthesis.

a. Explain why a shorter synthesis consisting of treatment of the starting material with sulfuric acid/nitric acid will not yield the desired target.

b. Is there a protecting group being used in this synthesis? If so, what is its identity, and what reaction is it preventing?

39. Design a synthesis of each target from the starting material given.

Read the assigned pages in your text, and do the assigned problems.

The Big Picture

A key skill in this ChemActivity is shuffling protons using curved arrows. In preparation for more complex mechanisms in upcoming ChemActivities, you should sit down with a blank piece of paper and make sure you can come up with a reasonable mechanism for imine, enamine, and acetal formation.

Another key objective of this ChemActivity is to refine your understanding of equilibrium and Le Châtelier's Principle. Most of the reactions in this part of the course are not exothermic/downhill, but approximately thermoneutral. Such reactions must be pushed to completion by adding a huge excess of reagents or, even better, by removing an auxiliary product (usually water).

Students tend to see reactions as either favorable or unfavorable; in this view, either a reaction happens and all reagents are consumed to produce products or no reaction occurs. This is a reasonable way to view a reaction that is steeply downhill or uphill in terms of energy. Such reactions (e.g. strong base + acid) were featured in the earlier sections of this book. Now we are encountering more and more reactions that are approximately thermoneutral. It does not make sense to think of such reactions as going 100% or 0%. The most likely equilibrium position for a thermoneutral step in a reaction is 50% products and 50% unreacted starting material. Fortunately, such reactions can be "pushed" one way or the other by manipulating other species that are found in the equilibrium expression. Most often this involves adding or removing water.

Common Points of Confusion

- The terms reversible and irreversible might more reasonably be replaced with "easily reversed" and "hard to reverse" since the terms reversible and irreversible imply absolutes that are not accurate. Reactions that are steeply downhill from left to right have an equilibrium that strongly favors the products. This means that it is nearly impossible to use Le Châtelier's principle to generate significant starting material from product. Reactions that are only slightly downhill from left to right are more easily reversed, but such grey area is discounted by the absolute terms reversible and irreversible.

- There are several alternate acceptable mechanisms for **imine**, **enamine**, and **acetal** formation. Students tend to show all of these, or none of these beginning with an initial protonation by an acid catalyst. In fact, the imine and enamine reactions are far more likely to begin with the nucleophile making a bond to the carbonyl carbon, while acetal formation is most likely to begin with acid protonation of the carbonyl.

  - The argument for why the imine/enamine reactions cannot start with protonation is that if the pH is low enough to protonate even 0.1% of the carbonyl oxygens in solution then it would also protonate 99.999+% of the amine molecules in solution (since, based on pK_a's, an amine is $\sim 10^9$ times more basic than a carbonyl oxygen). With almost no free amine the reaction would be so slow as to be impractical. These reactions must be run with just enough acid to protonate the oxygen to make it a leaving group.

  - The argument for acetal formation states that an alcohol is a poor nucleohile, and so acid is needed to protonate and activate the carbonyl electrophile. Adding enough acid to do this does not interfere with the nucleophilicity of the alcohol because an alcohol is a very weak base (comparable to the basicity of the carbonyl oxygen).

- Another point in these mechanisms where there is room for argument is in the proton transfers. Usually it is proposed that a solvent or other molecule picks up a proton from one atom and delivers it to another atom but you can also accomplish this via a cyclic transition state. Such **intramolecular proton transfer** steps do not involve any other molecule, and are most favorable for six-member ring transition states. As noted in the activity, four-member cyclic transition states are not generally favorable.

Notes

ChemActivity 8: Carboxylic Acids & Derivatives

(How does the acidic O–H group affect the reaction of a **carboxylic acid** with a nucleophile?)

Memorization Task 8.1: Recognize (and Name) Common Carboxylic Acid Derivatives

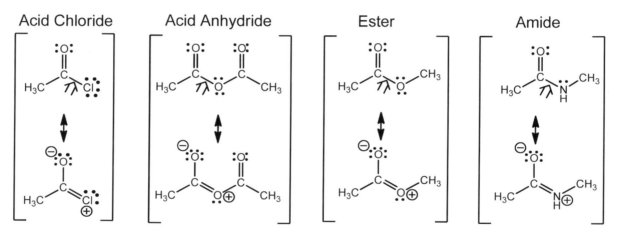

Nitriles (R-C≡N) are often considered another carboxylic acid derivative because a **nitrile** can be converted easily into a carboxylic acid by addition of acid and water.

Naming rules for carboxylic acids and derivatives are outlined in Nomenclature Worksheet 4.

Model 1: Carbonyl Reactivity

For each compound above we can draw a *second-order* resonance structure to show donation of electron density into the bond marked with =>. This reduces the reactivity of the carbonyl toward a nucleophile.

Construct Your Understanding Questions (to do in class)

1. For each set in Model 1, draw curved arrows showing conversion of the top resonance structure into the *second-order* resonance structure shown below it.

   a. π overlap is weak between atoms on different rows of the periodic table. Based on this, identify the one 2nd order resonance structure with a very weak π bond to carbon.

   b. (Check your work) Is your answer to the previous question consistent with the fact that the Cl atom in the acid chloride in Model 1 donates very little electron density to carbon?

   c. Of the other three, which one donates the <u>most</u> electron to carbon via a 2nd order resonance structure? (*Hint*: Consider electronegativity.) Explain your reasoning.

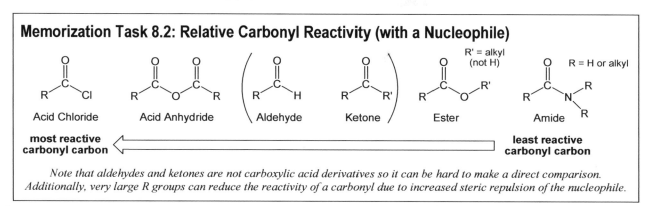

Memorization Task 8.2: Relative Carbonyl Reactivity (with a Nucleophile)

Acid Chloride   Acid Anhydride   Aldehyde   Ketone   Ester   Amide

most reactive carbonyl carbon   least reactive carbonyl carbon

Note that aldehydes and ketones are not carboxylic acid derivatives so it can be hard to make a direct comparison. Additionally, very large R groups can reduce the reactivity of a carbonyl due to increased steric repulsion of the nucleophile.

2. (Check your work.) Are your answers to Question 1 consistent with Memorization Task 8.2? Explain.

Model 2: Acid-Base Reactivity of a Carboxylic Acid (CA)

Construct Your Understanding Questions (to do in class)

3. Model 2 shows three possible ways that a nucleophile that is also a base (represented as Nuc $^{\ominus}$) could react with acetic acid. <u>Predict</u> which one of these pathways is favored over the other two.

    a. Draw the product(s) of each set of curved arrows, including resonance structures.

    b. Label each set of curved arrows with the appropriate reaction type. Choose from:
        Acid-Base, Elimination, Electrophilic Addition, or **Nucleophilic Addition**.

4. (Check your work.) Two of the three reactions depicted in Model 2 are acid-base reactions. Each of these acid-base reactions results in a conjugate base with two important resonance structures.

    a. Check that you drew <u>both important resonance structures</u> for each conjugate base.

    b. Circle the conjugate base that is lower in potential energy, and explain your reasoning. (Label this reaction in Model 2 "**favored reaction pathway.**")

5.   (Check your work.) Is your answer in part b in Question 4 consistent with Memorization Task 8.3?

Memorization Task 8.3: Carboxylic Acids Donate an H⁺ (except in acidic conditions)

A nucleophile that is also a base **will remove the acidic H from a carboxylic acid rather than bond to the carbonyl carbon**.

Carboxylic Acid   Carboxylate Anion

Once this H is removed, the anionic carboxylate will repel almost all nucleophiles.
Notable exceptions are $LiAlH_4$ (see Synth. Transf. 8.1), and Grignard/lithium reagents.

Synthetic Transformation 8.1: Carboxylic Acid Derivatives—Reduction with Hydride

Y = H (aldehyde) *
Y = Cl or Br (acid halide) *
Y = OCOR (acid anhydride) *
Y = OH (carboxylic acid)
Y = OR' (ester)

*will react with
$NaBH_4$

1) $LiAlH_4$ or $NaBH_4$
2) $H_3O^{\oplus}$ → primary alcohol

ketone *

1) $LiAlH_4$ or $NaBH_4$
2) $H_3O^{\oplus}$ → secondary alcohol

amide

1) $LiAlH_4$
2) $H_3O^{\oplus}$ → amine

6.   List functional groups that require use of the more powerful $LiAlH_4$ to accomplish a reduction.

Model 3: Nucleophilic Addition to a Carboxylic Acid (at low pH)

In acidic conditions (low pH) the O—H bond of a carboxylic acid stays intact. This allows nucleophilic addition to the carbonyl carbon by an alcohol to form an **ester**, as shown in Synth. Transf. 8.2.

Synthetic Transformation 8.2: Acid-Catalyzed Ester Synthesis

carboxylic acid   alcohol   acidic conditions   ester   boiling off water drives reaction to the right

Construct Your Understanding Questions (to do in class)

7.   Devise a mechanism for Synthetic Transformation 8.2. *Hint*: The first step is protonation of the carbonyl oxygen by the acid catalyst.

Extend Your Understanding Questions (to do in or out of class)

8. Construct an explanation for why the following amide-forming reaction analogous to the ester-forming reaction shown in Synthetic Transformation 8.2 does NOT produce the desired product. (*Hint*: Consider what species in solution is most likely to react with the acid catalyst.)

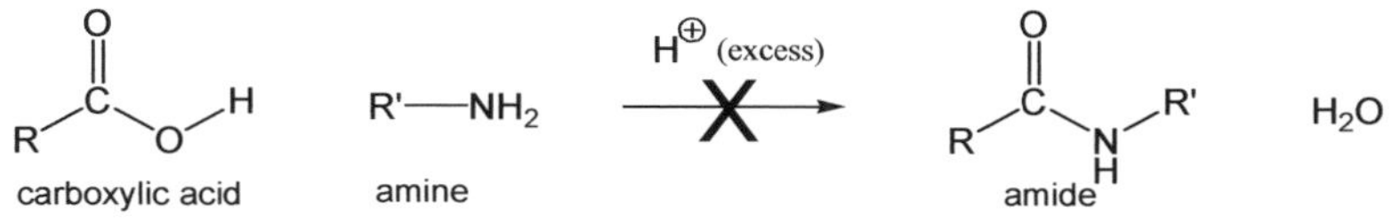

Memorization Task 8.4: Few Nucleophiles React Directly with the Carbonyl of a CA.

Alcohols are among the only nucleophiles that can undergo nucleophilic addition to the carbonyl of a **carboxylic acid**. Other, more basic nucleophiles fail because of one of the following reasons…

   (1) *(Under basic conditions—with excess basic nucleophile)* The nucleophile removes the acidic H from the carboxylic acid, rendering its carbonyl virtually unreactive.

   (2) *(Under acidic conditions—with excess acid)* A nucleophile that is also a good base will be protonated by the acid catalyst and consequently become non-nucleophilic

---

9. (Check your work.) Which reason in Memorization Task 8.4 best explains the failure of the reaction at the top of the page?

10. Only three of the following reactions produce a significant yield of the product shown.

   a. Cross out the reaction arrows of the other three reactions—the ones expected to fail.

   b. Put a (1) or (2) next to each failed reaction to indicate which reason in Memorization Task 8.4 best explains why that reaction is doomed to failure. (Species in *excess* are labeled)

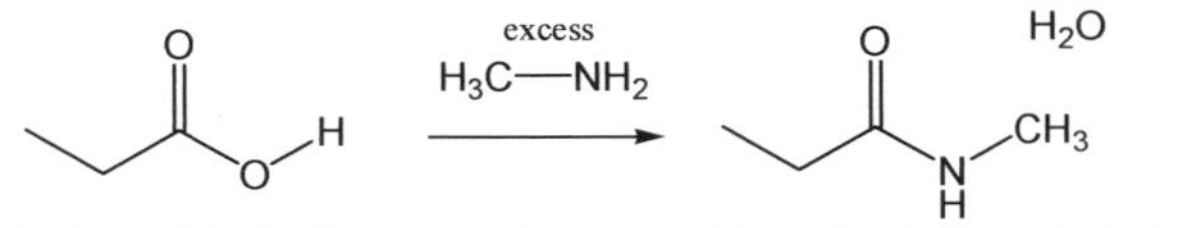

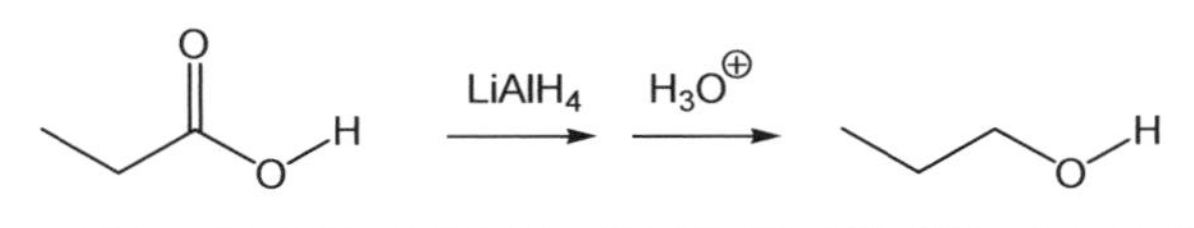

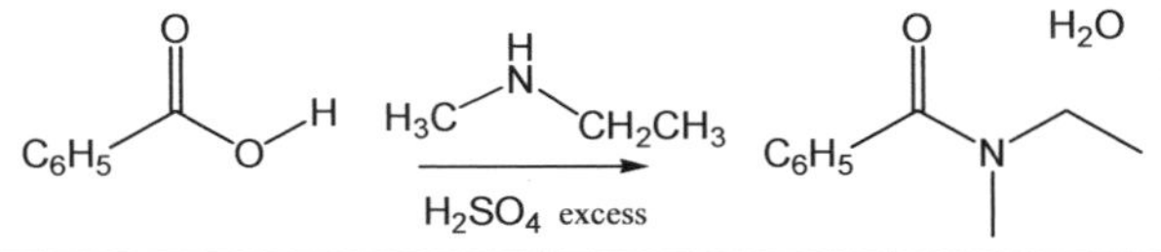

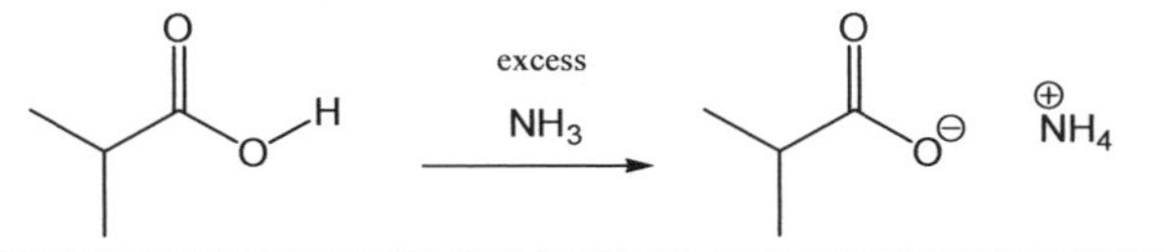

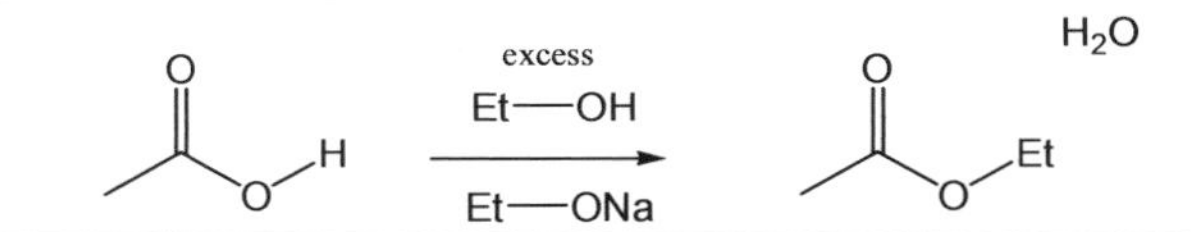

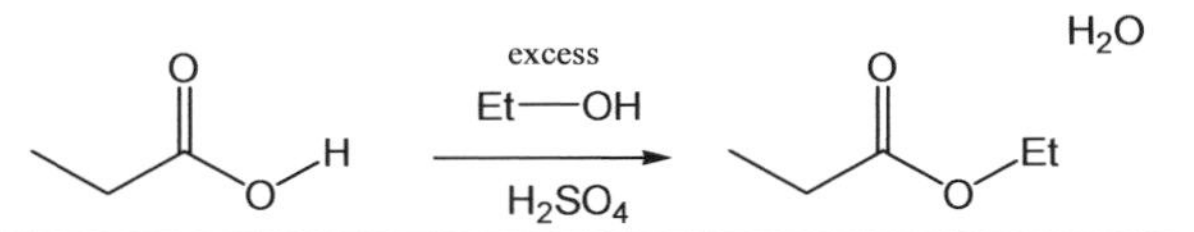

11. <u>Methoxide ion ($^-OCH_3$) is an excellent nucleophile, yet the following addition does not occur.</u>

a. Draw the reaction and resulting products that are likely to form instead of the one shown.

b. Construct an explanation for why the **conjugate base of acetic acid** is formed preferentially over the **nucleophilic addition product above**

12. Synthetic Transformation 8.2 is reversible. Describe a laboratory procedure that would allow you to reverse this reaction and generate a carboxylic acid from an ester as shown in Synthetic Transformation 8.3, below.

Synthetic Transformation 8.3: Acid-Catalyzed Ester Hydrolysis

13. Show the mechanism of Synthetic Transformation 8.3.

14. An attempt to generate a carboxylic acid from an ester under basic conditions leads to a carboxylate, as shown in Synthetic Transformation 8.4 below. (Note that the carboxylate can be easily protonated in a second step to give the desired carboxylic acid.) This reaction is called **saponification** for the *Latin* word for soap (*saponis)* because this reaction is used in the making of soap from fats (which are long-chain esters). Show the mechanism of this transformation.

Synthetic Transformation 8.4: Ester Saponification (Base-Catalyzed CA Synthesis)

15. Trans-esterification (the conversion of one ester into a different ester) can be accomplished using reaction conditions analogous to either Synth. Transf. 8.3 (acidic conditions) or Synth. Transf. 8.4 (basic conditions). Show the mechanisms of both acid- and base-catalyzed trans-esterification.

Synthetic Transformation 8.5: Trans-Esterification

16. Hydrolysis of an amide is similar to trans-esterification. Show the mechanisms of both acid- and base-catalyzed amide hydrolysis.

Synthetic Transformation 8.6: Acid- or Base-Catalyzed Amide Hydrolysis

17. Conversion of a carboxylic acid to an ester does NOT work in basic conditions, and the product in the box below does not form. Use curved arrows to show the mechanism and product of the side reaction that would result if this reaction were attempted in basic conditions.

18. The reaction in the previous question does not work in basic conditions. Explain how it is different from Synthetic Transformations 8.4 through 8.6, which do work in basic conditions.

Confirm Your Understanding Questions (to do in class)

19. Which compound in Model 1 has the most double-bond character in the bond marked in the top structure with an arrow =>? Explain your reasoning.

20. Construct an explanation for why a carbon NMR spectrum of N,N-dimethylformamide obtained at room temperature shows three peaks at 161 ppm, 36 ppm and 32 ppm. That is, construct an explanation for why the two methyl groups *not* equivalent.

21. Is your explanation in the previous question consistent with the finding that a carbon NMR spectrum of N,N-dimethylformamide obtained at an <u>elevated temperature</u> shows only two peaks: one at 161 ppm and the other at 34 ppm? Explain.

22. Label the O—H bond-breaking arrow in Model 2 with a "+5" to indicate that it takes approximately +5 pK_a units of energy to break this bond and liberate an H. (pK_a of a carboxylic acid is about 5.)

23. Label the C—H bond-breaking arrow in Model 2 with a "+20" to indicate that it takes approximately <u>+20 pK_a units of energy to break this bond and liberate the H on the carbon *alpha* (α) to the carbonyl.</u>
    Recall that normal C—H bonds take up to 50 pK_a units of energy to break. Resonance stabilization makes C_α–H_α bonds much easier to break. The resulting chemistry is the subject of upcoming ChemActivities.

24. Synthetic Transformations 8.7a/b are common ways of preparing a carboxylic acid. Use curved arrows to show the mechanism of each reaction. [Hint: 8.7b begins with electrophilic addition of acid and water to the triple bond of the nitrile.]

Synthetic Transformation 8.7a: 1° Grignard or Lithium Reagent with Carbon Dioxide

Synthetic Transformation 8.7b: Nitrile Hydrolysis

25. In many ways, the C≡N bond of a nitrile is analogous to the C=O bond of a carbonyl compound. Given this, devise a mechanism for the reaction at right. (*Hint*: An imine is one of the intermediates.)

Synth. Transf. 8.8: Grignard + Nitrile

26. Write appropriate reagents in the boxes over each reaction arrow.

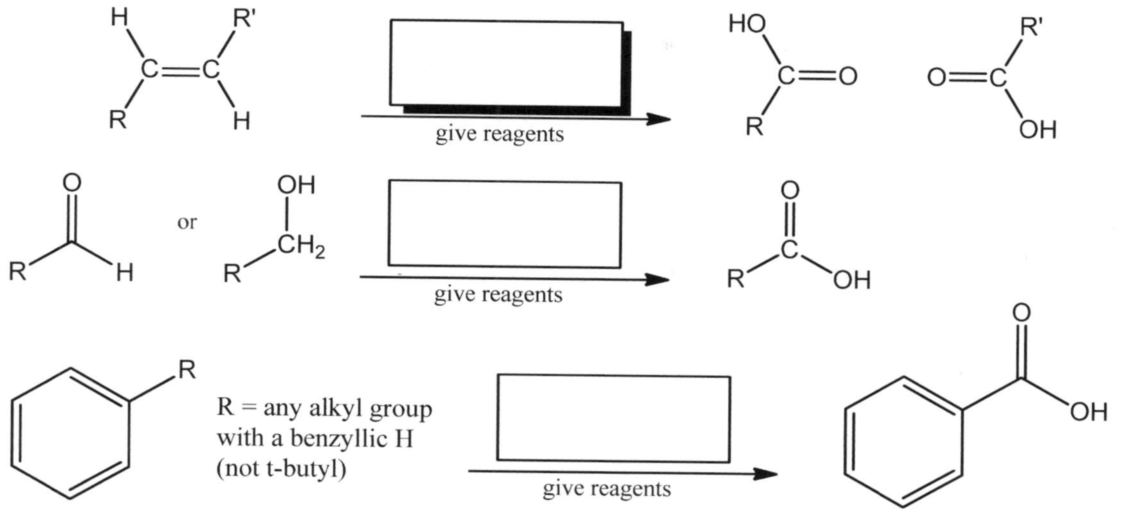

give reagents

give reagents

or

R = any alkyl group
with a benzyllic H
(not t-butyl)

give reagents

27. Devise a one-step mechanism for Synthetic Trasnformation 8.9. What type of mechanism is this?

Synthetic Transformation 8.9: Formation of an Ester from a 1° Alkyl Halide

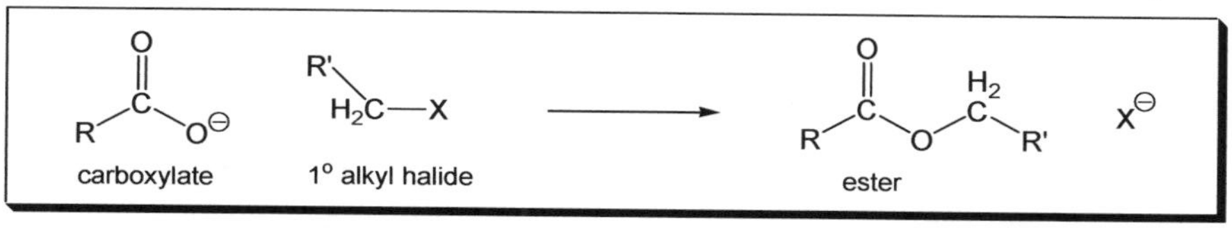

carboxylate   1° alkyl halide   ester

Read the assigned pages in your text, and do the assigned problems.

The Big Picture

In principle, carboxylic acids, esters, and amides can be interconverted by manipulating Le Cháelier's principle (as long as you avoid the complications caused by deprotonation of a carboxylic acid). Not all of these transformations are presented as official Synthetic Transformations because some are not often used in synthesis. For example, you can change an ester into an amide by treating it with excess amine and catalyst, and you can even do the opposite with sufficient time and heat. However, in practice it often makes sense in terms of time and overall yield to accomplish most carboxylic-acid-derivative syntheses by first making an activated carbonyl compound, either an acid halide or acid anhydride, both of which are covered in the next ChemActivity.

Common Points of Confusion

- It seems so simple, but many students forget that carboxylic acids are acidic! This gives rise to the following problems and resolutions…

    Nucleophilic addition to a **carboxylic acid** carbonyl group is usually impossible when the nucleophile is also a good base. (Nuc⁻ acts as a base and removes the acidic H^+ instead.)

    To solve this problem we must use a nucleophile that is a very weak base. For example, use methanol ($HOCH_3$) instead of methoxide ion ($^-OCH_3$).

    However, since methanol is a weak nucleophile, we must activate the carbonyl with an **acid catalyst to make the carbonyl carbon a better electrophile.**

- Remember that hydroysis of an ester or amide in <u>basic conditions</u> leads to a **carboxylate** (not a carboxylic acid).

Notes

ChemActivity 9: Acid Halides and Anhydrides

(Is there a way to make an amide from a carboxylic acid?)

Model 1: Acid Halides (Alkanoyl Halides)

A carboxylic acid can be transformed into an acid halide via Synthetic Transformation 9.1. The mechanism of this reaction is beyond the scope of this activity.

Synthetic Transformation 9.1: Preparation of Acid Halides from Carboxylic Acids

Acid halides are extremely versatile precursors for a wide range of carboxylic acid derivatives.

Synthetic Transformations 9.2a-i: Reactions of Acid Halides

Construct Your Understanding Questions (to do in class)

1. (Review) Explain *why* the following transformation fails to give the product shown. What product is formed instead?

2. Devise a synthesis of ethyl acetamide from acetic acid using the transformations in Model 1.

3. Explain how use of an acid halide circumvents problems you cited in your answer to Question 1.

4. Shown below is another synthesis of a carboxylic acid derivative (an ester) via an acid chloride. Use curved arrows to devise a mechanism of the <u>second part</u> of this synthesis (**Step B**).

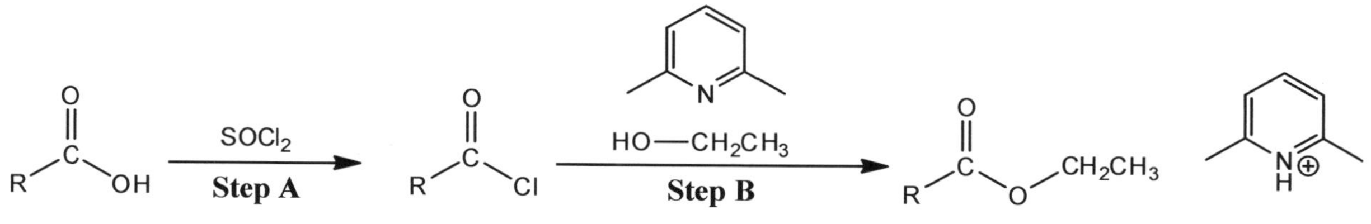

Memorization Task 9.1: S_N1 and S_N2 reactions cannot take place at *sp²*-hybridized carbons

5. (Check your work) Many students give the following ***partially incorrect*** answer to Question 4.

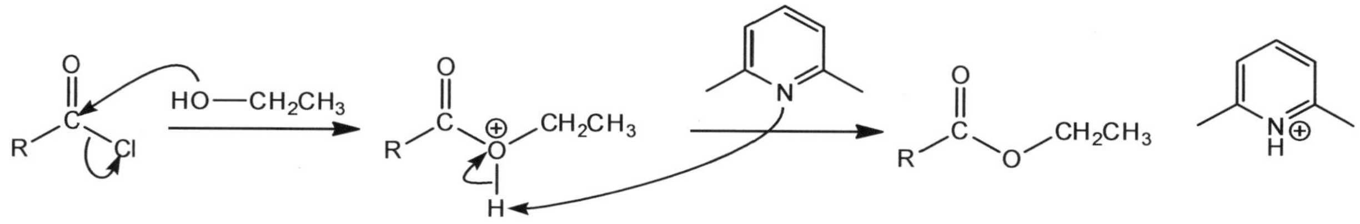

   a. Does this answer violate Memorization Task 9.1? If so, how.

   b. Does your mechanism at the top of the page violate Memorization Task 9.1? If so, devise a new mechanism. *Hint: it should start with a (two-step) nucleophilic addition-elimination, and include the following tetrahedral intermediate.*

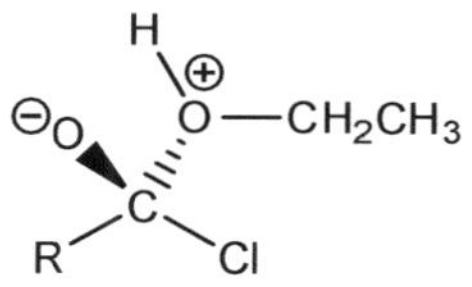

6. When excess hydride (e.g., $LiAlH_4$), lithium reagent (R-Li), or Grignard (R-MgBr) is added to an acid halide, **two molar equivalents** of nucleophile are incorporated into the product. Devise a mechanism for this reaction, assuming the nucleophile in this case is hydride anion (H^-).

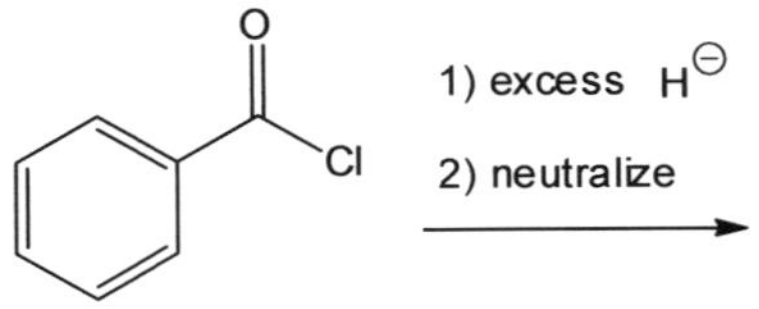

Model 2: Carboxylic Acid Anhydrides

Reactions with acid halides are often exothermic and rapid, leading to unwanted side products. Acid anhydrides are used in place of acid halides when a slower, more controlled reaction is desired.

*The name **acid anhydride** refers to the fact that you can generate a symmetrical acid anhydride by heating a pure sample of the corresponding carboxylic acid while removing the water that is produced ("anhydride" means "without water"). An easier way to generate an acid anhydride is to mix an acid halide with a carboxylate anion, as shown below.*

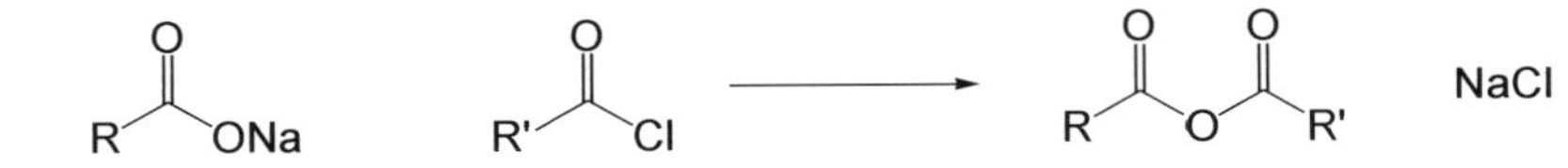

carboxylate ion **acid chloride** **acid anhydride**

Construct Your Understanding Questions (to do in class)

7. (E)According to Model 2, which is more reactive, an **acid halide** or **acid anhydride**? [circle one]

8. Devise a mechanism for the reaction in Model 2.

9. The ester synthesis on the previous page can be accomplished via an acid anhydride as shown below. Show the mechanism of **Step C** in this synthesis. (Another product is formed in this reaction, but is not shown below. What is it?)

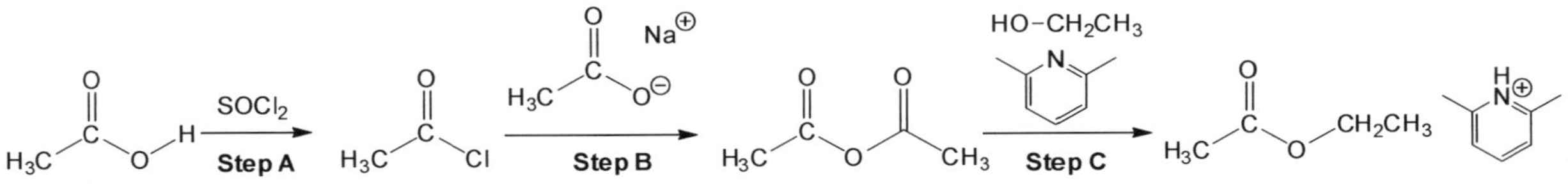

Model 3: Esters as Replacements for Acid Halides/Anhydrides

One explanation for why an acid anhydride is less reactive than an acid halide is that a carboxylate group (in the dotted box below, left) is a worse leaving group than chloride or bromide.

You can think of an ester as an even less reactive replacement for an acid halide or anhydride.

ester   acid anhydride   acid halide   X = Cl or Br

Extend Your Understanding Questions (to do in or out of class)

10. Circle the part of the ester in Model 3 that will act as a leaving group in an addition-elimination reaction. Construct an explanation for why esters are <u>less reactive</u> than acid anhydrides or acid halides in addition-elimination reactions.

Mem. Task 9.2: OR can <u>only</u> serve as a leaving group in strongly basic conditions

Strongly basic conditions, such as those generated by lithium reagents and Grignard reagents (but not lithium dialkyl cuprate reagents), or $LiAlH_4$ (but not $NaBH_4$) allow the breaking of a C—OR bond to form an alkoxide ion.

11. Devise a mechanism for the following transformation.

1) excess CH_3Li
2) neutralize

12. On the diagram below (from Model 1) an acid anhydride has been added to the starting box since reactions **a-i** all work with either an acid halide or an acid anhydride. Identify the letter of each reaction that works with an ester.

Synthetic Transformations 9.2a-i: Reactions of Acid Halides (adapted from Model 1)

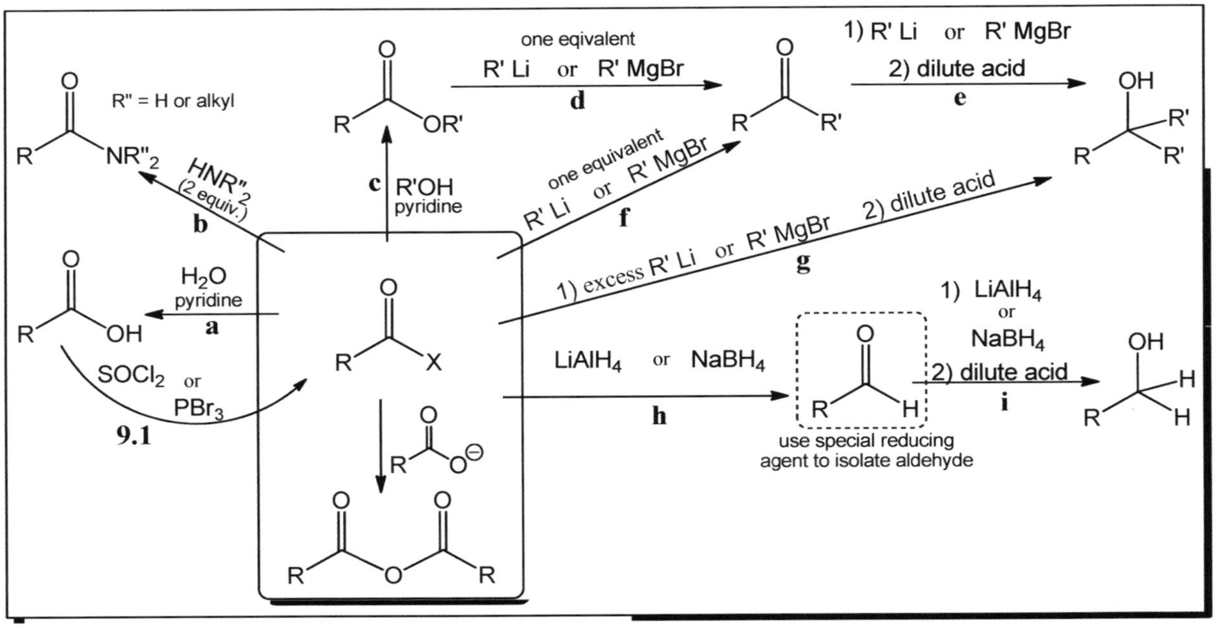

13. Label the structures in the Synthetic Transformations above according to their functional group category. Choose from **alcohol** (specify 1°, 2°, or 3° if indicated), **amine, aldehyde, ketone, amide, ester, carboxylic acid, carboxylate, acid halide,** or **acid anhydride**.

Confirm Your Understanding Questions (to do at home)

14. This reaction from Question 4 (copied below) requires addition of base. In this case 2,6-dimethylpyridine (2,6-lutidine) is used.

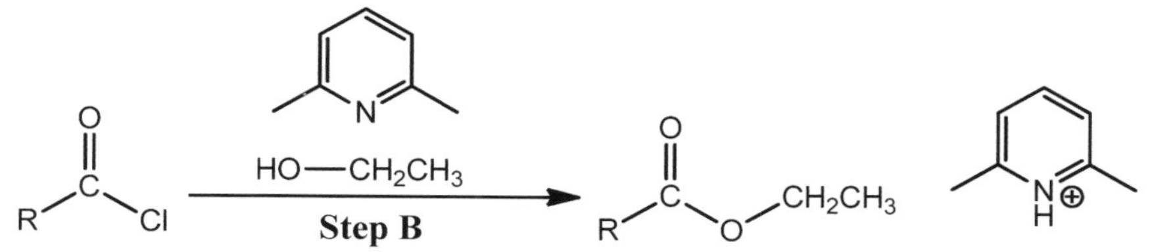

a. Construct an explanation for why 2,6-dimethylpyridine is better than pyridine. (Hint: what type of side product is made less likely by the two methyl groups?)

b. In Question 2 (copied below) no auxiliary base is needed, but it does require excess amine. Show the mechanism of this reaction and explain why two molar equivalents of base are needed. What other product is formed but not shown below?

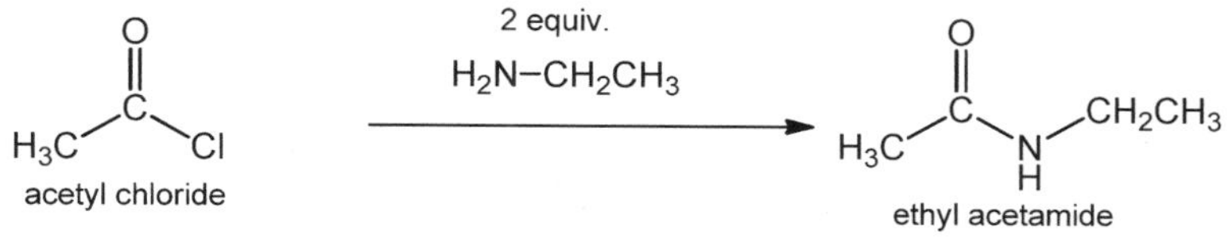

15. Draw the major product/s of each reaction at right →

Write "**no reaction**" if you do not expect a reaction.

Assume…

- nucleophile is present in <u>excess</u>

and that…

- if necessary, the product mixture is <u>neutralized</u> at the end of each reaction.

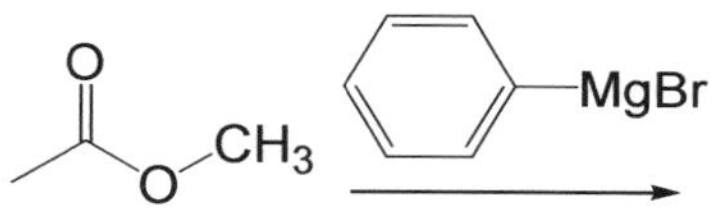

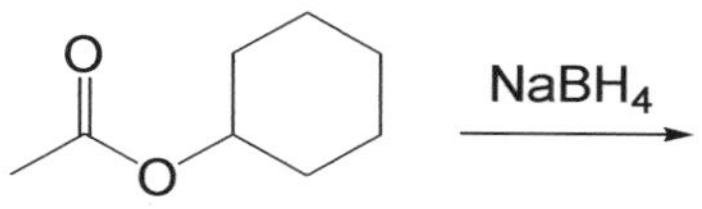

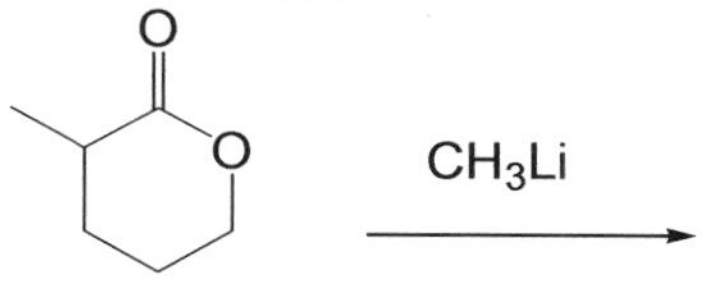

example of a lactone

16. (Check your work.) A cyclic ester is called a **lactone**. The reaction of the lactone in the third reaction in the previous question yields **4,5-dimethylhexane-1,5-diol**. Draw this product, check your work above, and devise a mechanism for this reaction (assuming the nucleophile is a H_3C^- anion).

17. The acid anhydride shown in Model 2 has an R group on one end and an R' group on the other end. When R = R' this it is called a **symmetrical acid anhydride**. (These are most common.) When R and R' are different it is called an **asymmetrical acid anhydride**.

    a. (E) Identify each acid anhydride below as symmetrical or asymmetrical.

    b. Construct an explanation for why use of an asymmetrical acid anhydride usually gives rise to two different products.

    c. Draw both amide products that arise if the asymmetrical acid anhydride below is treated with excess ammonia (NH_3).

18. Rank the following compounds from…"1 = best electrophile" to "4 = worst electrophile," and explain your reasoning.

19. For each reaction, state which product is more likely, and explain your reasoning.

20. Draw the major product, and show the mechanism of each of the following reactions with an acid catalyst and then with an appropriate base catalyst. (Draw the structure of the best base catalyst for each reaction.)

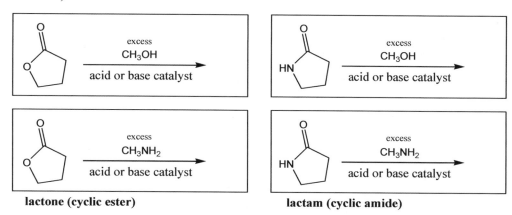

lactone (cyclic ester)   **lactam (cyclic amide)**

21. Try to draw the most likely product of each of the following reactions, then check your work by drawing the mechanism of the reaction. *Tip: These types of questions can be very difficult to answer correctly without sketching the mechanism.*

22. Complete the synthesis of the asymmetrical anhydride, and draw all four organic products that result (two esters and two carboxylic acids).

23. Acid halides and acid anhydrides allow convenient synthesis of carboxylic acid derivatives because they each have a good leaving group attached to the carbonyl carbon.

    a.   What is this leaving group in the case of an acid halide?

    b.   What is this leaving group in the case of an acid anhydride? …an ester?

    c.   If you use a nucleophile that is strong enough, you don't need a good leaving group. A Grignard reagent is an example of such a nucleophile. In the following synthesis of an alkoxide. Show the mechanism of the reaction, assuming the nucleophile is $CH_3CH_2^-$. (Remember that OH or OR are good leaving groups ONLY in strongly basic conditions, such as in a Grignard or Lithium reaction.)

$$H_3CH_2C-MgBr$$

24. Repeat part c of the previous question using $LiAlH_4$ in place of ethyl magnesium bromide.

Note that organometallic reagents (lithium reagents, Grignard reagents, $LiAlH_4$, etc.) often have a mechanisms of action that involve the metal atom and are more complicated than if the nucleophile were truly a carbanion or a hydride with a simple counterion. For our purposes the correct product and a reasonable answer can be found by assuming these reagents work as if they were a carbanion or a hydride with a simple counterion.

25. Starting from any carboxylic acid(s), devise a synthesis of the 1st three acid anhydrides in Question 17a.

26. The R* on the molecules below represents a chiral R group. Assume that the carboxylic acid starting material is very costly because of this chiral group. Construct an explanation for why, in this case, it would be a poor choice to use an acid anhydride to transform this acid into a derivative (e.g., the dimethyl amide shown). *Hint*: Draw the other organic product produced in Step C.

27. Construct an explanation for why the following asymmetrical anhydride gives a much higher yield of ethyl acetate than ethyl 2,2-dimethylpropionate.

28. You may have encountered lithium dialkyl cuprate reagents in your course. These reagents (sometimes abbreviated LiR$_2$Cu) deliver a carbon nucleophile that is comparable to a Grignard or lithium reagent. However, as compared to Grignard (R-MgBr) and ordinary lithium (R-Li) reagents, LiR$_2$Cu reagents are much…

- <u>less</u> reactive toward carbonyls (C=O)
    - o LiR$_2$Cu only react with acid halide/anhydride (<u>not ester, aldehyde or ketone, etc.</u>)
- <u>more</u> reactive toward alkyl halides (R-X)
    - o LiR$_2$Cu were developed to fill the gap caused by the fact that Grignard/lithium reagents do not give good yields in reactions with alkyl halides

Based on the information above, draw the most likely product of each reaction.

- Write "**no reaction**" if you do not expect a reaction.
- Assume the nucleophile is present in <u>excess</u> (except where noted)
- Assume that, if necessary, the product mixture is neutralized at the end the reaction

29. Design an efficient synthesis of each of the following target molecules. All carbon atoms must come from the starting material(s) given.

Target			
Starting material			

Read the assigned pages in your text, and do the assigned problems.

The Big Picture

In principle, carboxylic acids, esters, and amides can be interconverted by manipulating Le Cháelier's principle (as long as you avoid the complications caused by deprotonation of a carboxylic acid). However, in practice it often makes sense in terms of time and overall yield to accomplish most carboxylic-acid-derivative syntheses by first making an activated carbonyl compound, either an acid halide or acid anhydride (or ester).

Common Points of Confusion

- Students often forget that use of excess $LiAlH_4$, lithium, or Grignard reagent with an acid halide, acid anhydride or ester will result in incorporation of <u>two</u> molar equivalents of nucleophile in the product; whereas lithium dialkyl cuprates will deliver only one equivalent to an acid halide or anhydride (since lithium dialkyl cuprates do NOT react with aldehydes, ketones, esters, etc.).

- Reduction of an **ester** (RCO_2R') with a **hydride** reagent (e.g., $LiAlH_4$) leads to an alcohol (ROH), not an ether (ROR') as some students assume. The source of this misconception appears to be the fact that reduction of a carboxylic acid or amide with this same reagent appears to imply that the reagent turns a C=O into a CH_2. The mechanism of the ester reduction reveals that the OR group is lost as a leaving group. (This is one of the rare instances of RO^- functioning as a leaving group). The carbonyl oxygen remains and becomes the alcohol oxygen of the product.

- Synthetic transformations on test and quizzes often do not specify how much of a particular reagent is added. In such situations you are to assume that the appropriate amount is added. That is, if the reagent is a catalyst that is required only in trace amounts, then assume there is a trace amount (even if it doesn't say "trace amount" next to the reagent). Conversely, if excess reagent is necessary to effect a reasonable reaction it may not say "excess" next to the reagent, but you are to assume excess reagent is present. The exception to this rule is when 1 equivalent of a reagent gives a very different product than multiple equivalents. In such instances (such as the addition of 1 molar equivalent of a bulky Grignard to an acid halide, anhydride or ester at low temperature) the number of molar equivalents of reagent will likely be specified.

- Ring opening of lactones (cyclic ester) and **lactams** (cyclic amides) could have been included in the previous activity (but there were already too many long mechanisms there, so they are presented here). The key error that students make in answering questions (especially multiple-choice questions) is to assume that the ring oxygen of a **lactone** remains attached to the carbonyl carbon in a resulting ester or carboxylic acid. This misconception is often tested using an O^{18}-labeled oxygen as in Question 21. The best way to answer such questions is to quickly sketch the mechanism of the reaction.

Notes

ChemActivity 10: Enolate and Enol Nucleophiles

(How can an aldehyde or ketone act as a nucleophile?)

Model 1: Acid-Base Reactivity of Aldehydes

Many nucleophiles are also good bases (e.g., RO^-, R_2N^-, R_3C^-, H^-)

In the past few ChemActivities we have focused on reactions such as the one at right, in which a basic species <u>acts as a nucleophile</u>, making a bond to a carbonyl carbon.

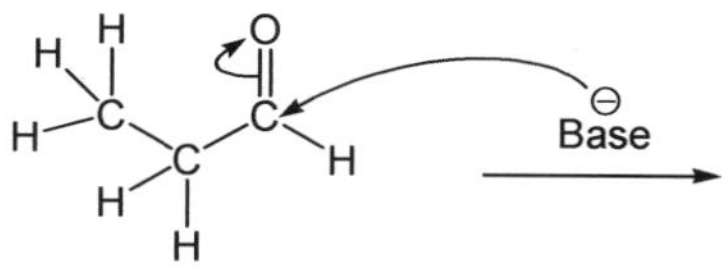

A nucleophile is often also a base. Shown below are three possible acid-base reactions of this pair.

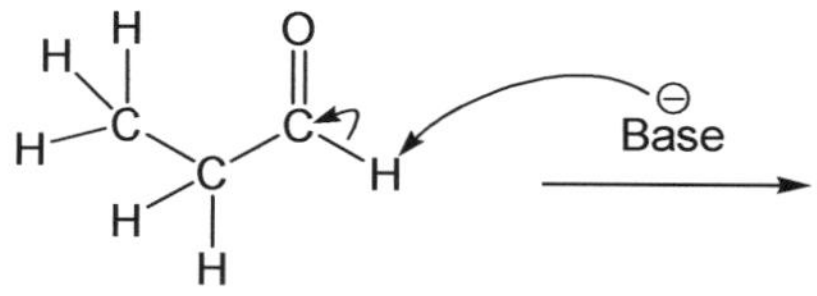

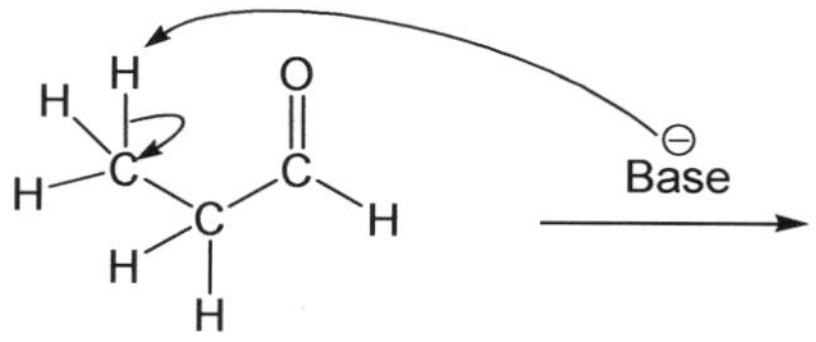

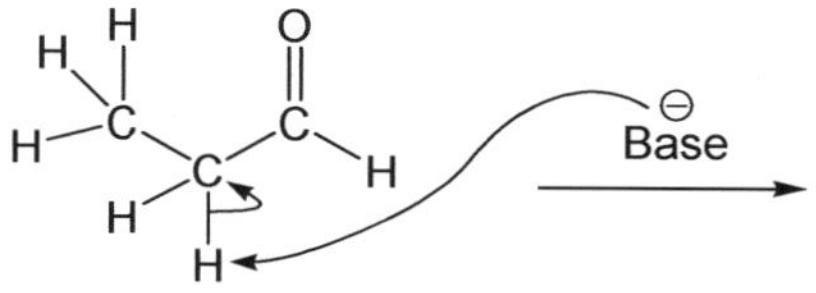

Construct Your Understanding Questions (to do in class)

1. Draw the product of each set of curved arrows. (Include all important resonance structures.)
2. Which one of these three acid-base reactions is most favorable? Explain your reasoning.

3. (Check your work.) Is your answer above consistent with Memorization Task 10.1?

Memorization Task 10.1: Enolate Ions

The most acidic H on a carbonyl compound (except a carboxylic acid) is the **alpha hydrogen**. Removal of an H_α gives a resonance-stabilized **enolate ion** that acts as a versatile carbon nucleophile.

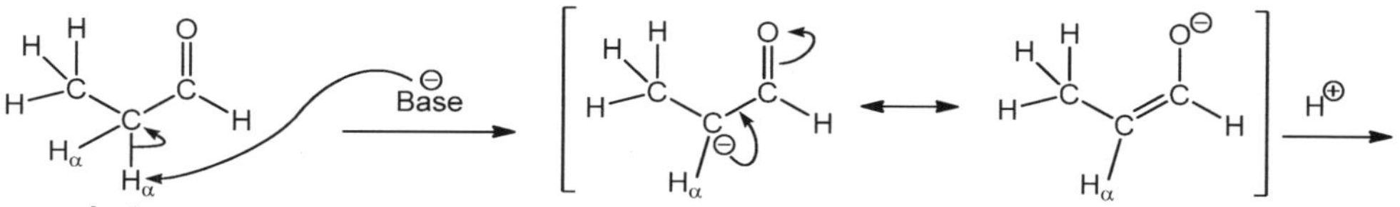

keto   enolate   enol

---

4. Draw the structure of the enol that is missing from Memorization Task 10.1.

Model 2: Enolate as Nucleophile

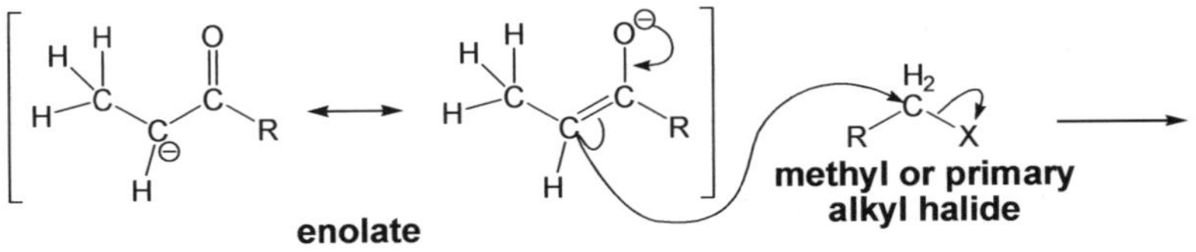

enolate

methyl or primary
alkyl halide

Construct Your Understanding Questions (to do in class)

5. Either resonance form of an enolate can be shown making a bond to an electrophile. A very useful application of this, an enolate in reaction with a methyl or primary alkyl halide, is shown above.

   a. Draw the product that results from the curved arrows shown in Model 2.

   b. Show this same reaction using the other enolate resonance structure (redrawn below).

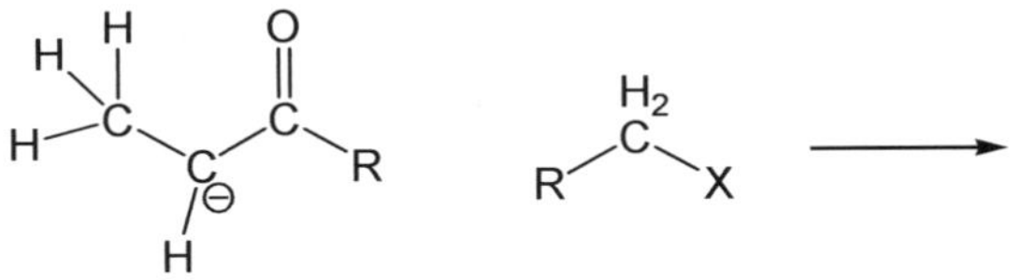

   c. What is the name of this type of mechanism?

Synthetic Transformation 10.1: Base-Promoted α-Alkylation of Carbonyl Compounds

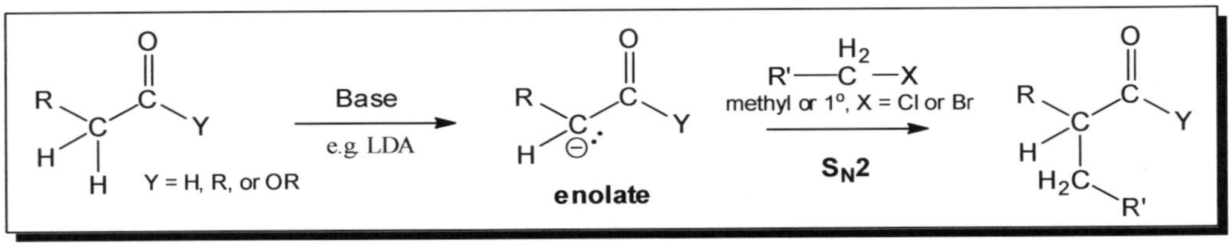

6. Do you expect **LDA** (structure shown at right) to be a strong base, a good nucleophile, or both? Explain your reasoning.

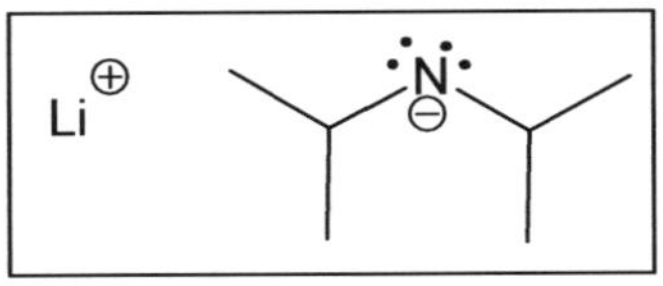

   Lithium Di-isopropyl Amide

   "LDA"

7. (Check your work.) Is your answer above consistent with the fact that LDA is an excellent base for Synthetic Transformation 10.1 because it produces almost exclusively enolate (and rarely bonds to the carbonyl carbon)?

Model 3: Problems with Base-Promoted α-Halogenation

In principle, it should be possible to replace one H$_\alpha$ with a halogen as shown below…

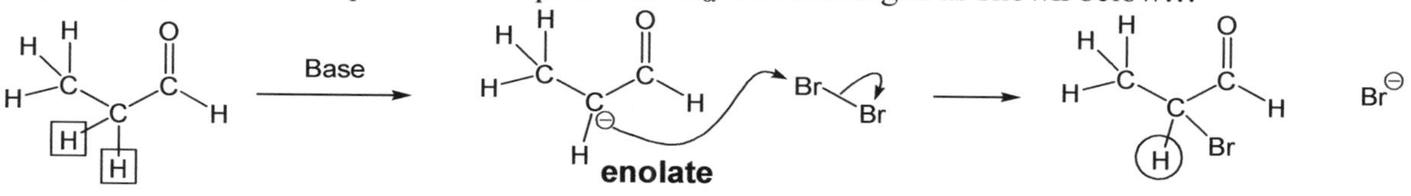

Unfortunately, the product above is more acidic than the starting material so the 2nd H$_\alpha$ is also replaced…

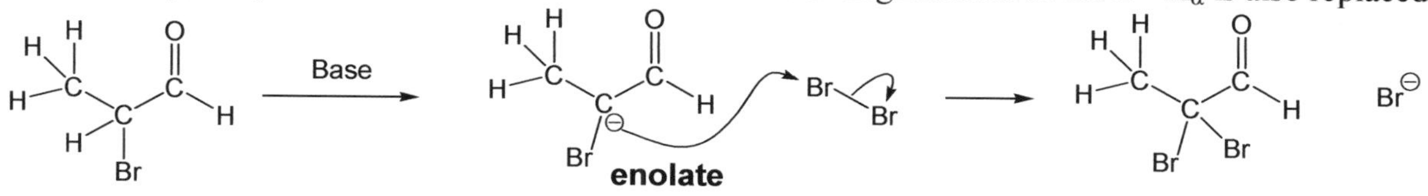

This reaction cannot be controlled and stops only when all alpha hydrogens have been replaced.
*Base-promoted halogenation of <u>methyl</u> ketones leads to an interesting result known as the **haloform reaction** (See homework.).*

Construct Your Understanding Questions (to do in class)

8. Construct an explanation for why the circled H in Model 3 is more acidic than the boxed H's.

Model 4: Acid-Catalyzed α-Halogenation (Enol Nucleophiles)

<u>Mono</u>-α-halogenation can be achieved using acidic (instead of basic) conditions. In acidic conditions the active nucleophile is an enol instead of an enolate.

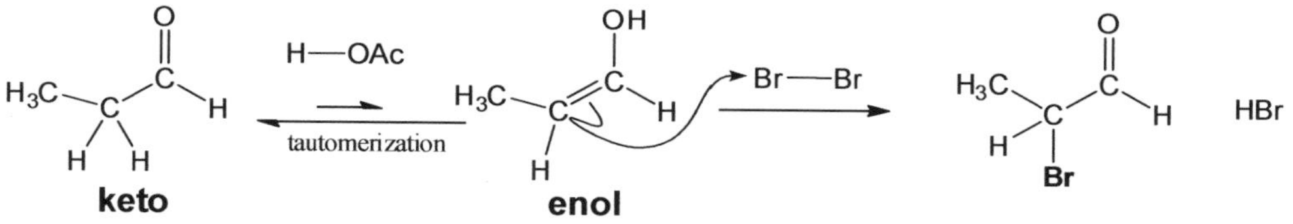

Construct Your Understanding Questions (to do in class)

9. Devise a mechanism for the (overall four-step) reaction in Model 4. The active nucleophile in acid (the enol) and one curved arrow (depicting formation of the C—Br bond) are shown for you.

Synthetic Transformation 10.2: Acid-Catalyzed α-Halogenation of Carbonyl Compounds

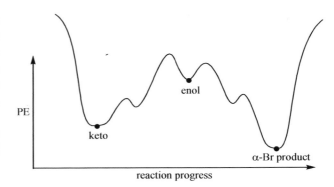

Extend Your Understanding Questions (to do in or out of class)

10. In a basic solution of an alcohol, alkoxide (RO⁻) can act as a nucleophile, but in an acidic solution of an alcohol RO⁻ is NOT available; the best nucleophile available for reactions is therefore ROH.

   a. Explain why an alcoxide (RO⁻) is NOT found in an acidic solution (*except in trace amounts*).

   b. What is the best nucleophile available for reaction...

       (1)   in a **basic** solution of <u>aldehyde</u>?

       (2)   in an **acidic** solution of <u>aldehyde</u>? (Check your work: See Models 3 and 4.)

   c. Which is a better nucleophile **RO⁻** or **ROH**? (circle one)

   d. Which is a better nucleophile **enolate** or **enol**? (circle one)

   e. (Check your work.) Is your answer above consistent with the fact that acid-catalyzed α-halogenations are much slower and more controlled than those in basic conditions?

11. Assume 99% of the starting aldehyde in Model 4 is in the keto form. Explain how most of the propanal can be converted to 2-bromopropanal even though, at any given moment, **less than 1% of the propanal in solution is in the enol form.**

   In other words, what happens once all the propanal that was originally in the enol form has been converted to product?

Model 5: pK_a Values of Selected Carbonyl Compounds

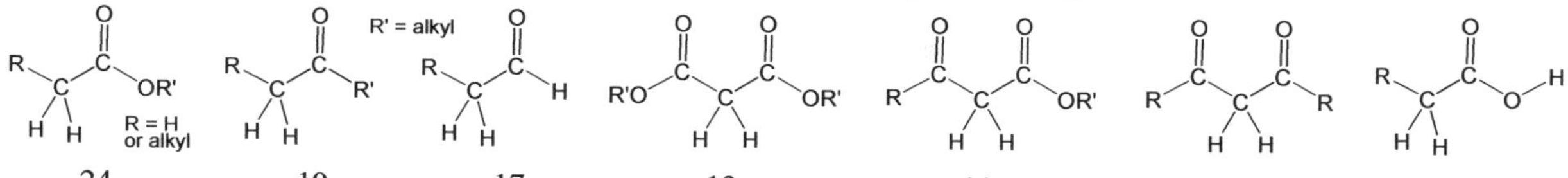

24	19	17	13	11	9	5

Extend Your Understanding Questions (to do in or out of class)

12. Identify any **beta-keto carbonyl compounds** (β-keto carbonyl compounds) in Model 5. (Recall from Nomenclature Worksheet 4 that a β-keto carbonyl compound has a C=O on the "beta carbon" of a carbonyl functional group.)

13. Consider the pK_a data in Model 5.

    a. For each compound, circle the H or H's associated with the pK_a value listed below it.

    b. Which is more acidic, **an aldehyde** or **a ketone**? (circle one)

    c. Which is more acidic, **a ketone** or **an ester**? (circle one)

14. Draw the conjugate base of the β-keto ketone below, including all important resonance structures.

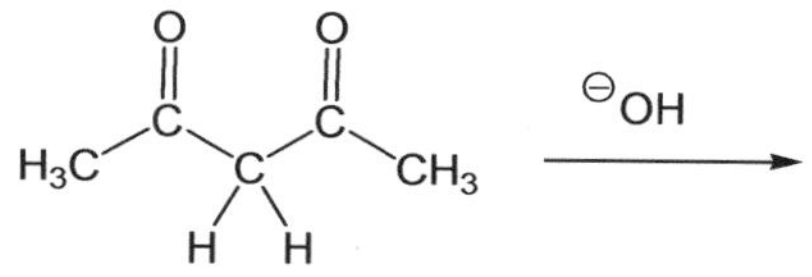

15. Construct an explanation for why the β-keto ketone above (with a pK_a of about 9) is much more acidic than an ordinary ketone (with a pK_a of about 19).

16. Draw the most likely product of the following reaction sequence:

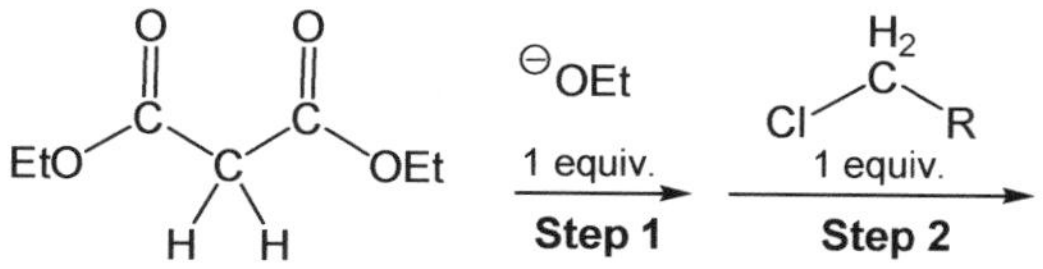

17. The previous question shows the first two steps in a common synthesis of carboxylic acids starting from a β-keto diester (also called a **malonic ester**).

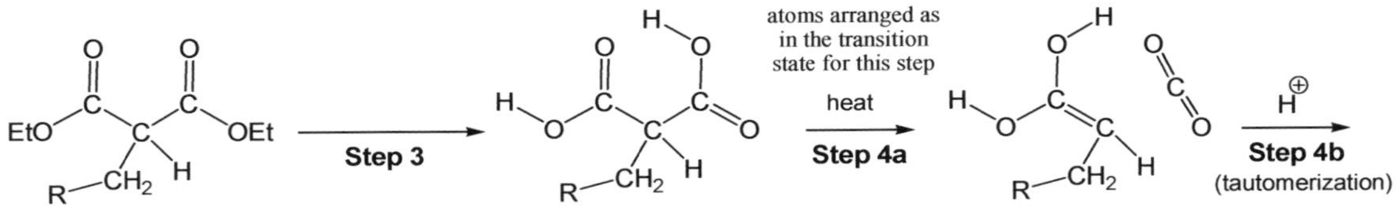

    a.   Fill in appropriate reagents over the reaction arrow for Step 3.

    b.   Devise a one-step mechanism for Step 4a (called **decarboxylation** = loss of CO_2).

    c.   Draw the final carboxylic acid product (the keto tautomer of the enol shown above).

Synthetic Transformation 10.3: Malonic Ester Synthesis of Carboxylic Acids

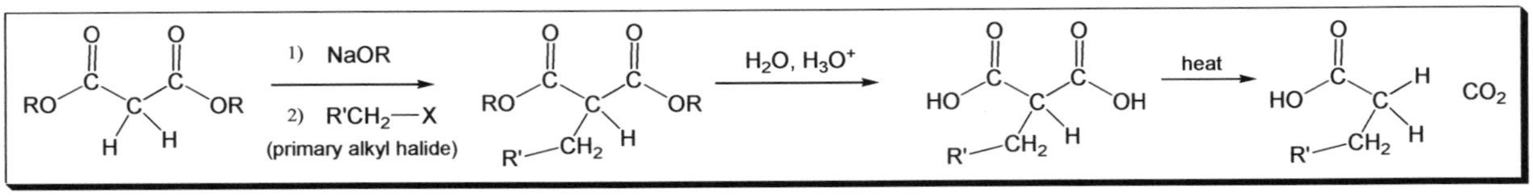

Confirm Your Understanding Questions (to do at home)

18. Construct an explanation for why it is much easier to remove an H from a carboxylic acid than from an aldehyde.

19. Draw the products of Rxn's A and B, below.

1) KOEt
2) I—CH₃
Rxn A

1) KOEt
2) I—CH₃
Rxn B

20. Explain why the examples below are called **β-keto** aldehydes or ketones.

21. The β-keto aldehyde on the far left, above, has three different possible conjugate bases.

   a. Draw all three.

   b. One conjugate base is not resonance-stabilized at all. Which one is this?

   c. Of the two resonance-stabilized conjugate bases, one has a total of two resonance structures, and the other has a total of three. Draw all resonance structures of both.

   d. Draw in the most acidic H (or H's) on each **β-keto** compound in the previous question.

22. Explain why LDA is used in Synthetic Transformation 10.1, but is not needed in Synthetic Transformation 10.3. In other words, explain why a weaker alkoxide base is sufficient to effect the latter transformation.

23. Describe how an acid anhydride differs from a β-keto carbonyl compound (such as a malonic ester).

24. Which reaction below is faster, Step 1 or Step 2? Explain your reasoning. (Assume that in each reaction the rate-limiting step is removal of the H by hydroxide.)

NaOH/Br₂
Step 1

NaOH/Br₂
Step 2

25. Is your answer above consistent with the fact that this type of base-promoted α-halogenation is very difficult to control and usually results in multiple α-H's being replaced with halogen atoms? Explain.

26. Base-promoted halogenation of a **methyl** ketone (or aldehyde) gives rise to the following (somewhat unexpected) product. Devise a mechanism for the portion of this reaction in the box.

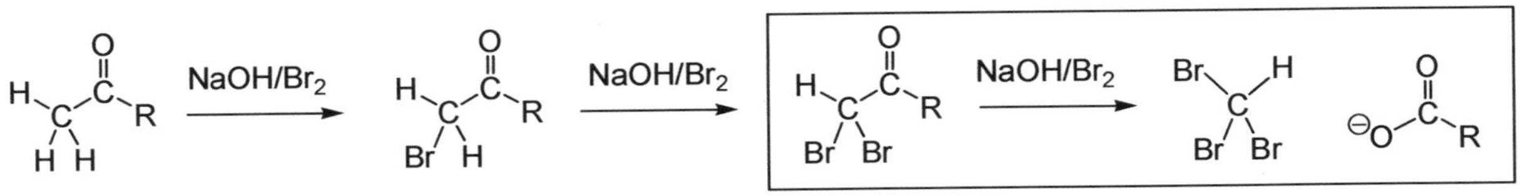

Synthetic Transformation 10.4: Haloform Reaction (and Iodoform Test for Methyl Ketones)

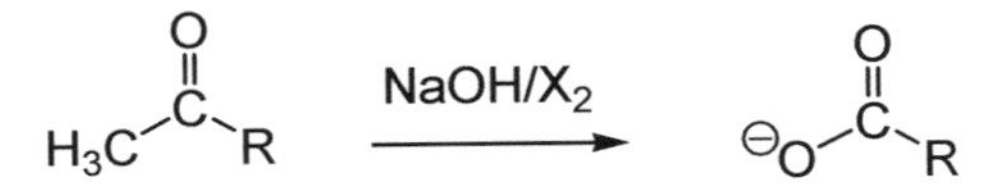

methyl ketone

X_3CH

haloform

When X = I, the product (I_3CH) is a yellow precipitate. This can be used as an analytical test for the presence of a methyl ketone.

27. Based on the energy diagram in Question 11, which is rate-limiting in acid-catalyzed alpha-halogenation, **keto-enol tautomerization** or **halogenation**? (circle one)

28. Assuming your answer above holds true for all acid-catalyzed alpha substitution reactions, which of the following acid-catalyzed reactions (if either) is expected to be faster?
alpha-chlorination or **alpha-bromination** or **neither (rates are the same)** (circle one)

29. Use curved arrows to show the mechanism of an acid-catalyzed tautomerization.

30. Show the mechanism of the following reaction:

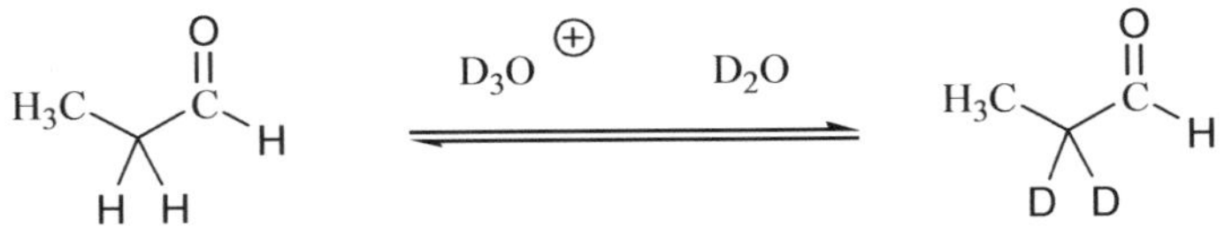

31. The reaction in Model 3 is called base-***promoted*** halogenation, while the reaction in Model 4 is called acid-***catalyzed*** halogenation. Explain the use of the different terms in quotes in each case.

32. A solution of pure **R** chiral aldehyde (shown below) was accidentally exposed to a small amount of acid. The next day, the solution was found to contain a racemic mixture (1:1 mix of **R** and **S**). Draw a mechanism to show how conversion of R into S could have occurred.

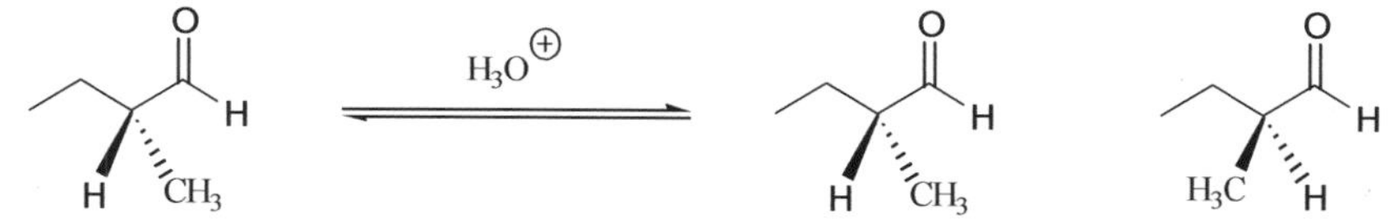

pure R stereoisomer   racemic mixture (1:1)

33. Draw the keto form of phenol. Explain why phenol is more stable in its enol form. (Note that phenol is one of the very few aldehydes or ketones for which the enol form is favored.)

34. Both enols and enolates can act as nucleophiles. Which one is a better nucleophile? Under what circumstances does each act as a nucleophile?

35. Acid-catalyzed alpha halogenation does not work for esters, amides, or carboxylic acids, <u>but it does work for acid halides</u>. This fact can be exploited to generate α-brominated carboxylic acid derivatives as shown in Synthetic Transformation 10.5.

Synthetic Transformation 10.5: α-Bromination of Carboxylic Acids (Hell-Vollhard-Zelinsky)

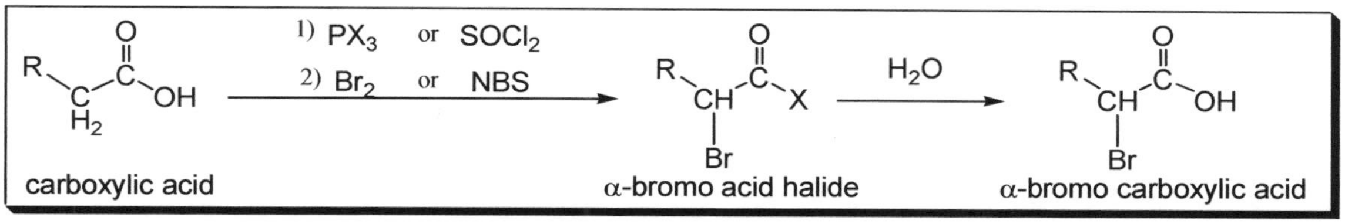

carboxylic acid   α-bromo acid halide   α-bromo carboxylic acid

   a.   Show the mechanisms of acid-catalyzed bromination of an <u>acid chloride</u> (Assume the reagents are Br_2 and H^+, though the catalyst in the reaction above is a phosphorus species.).

   b.   What reagents could be used in place of water in the last step of Synthetic Transformation 10.5 to generate a methyl ester? …an *N,N*-dimethyl amide?

36. Alpha-bromination of a carboxylic acid is a critical step in some laboratory syntheses of amino acids. Devise a synthesis of <u>racemic</u> alanine from propionic acid.

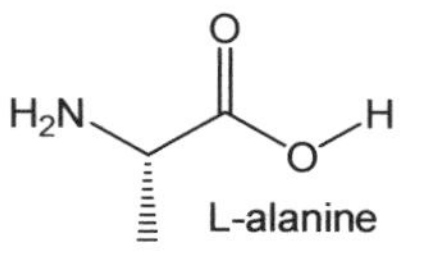

L-alanine

37. Explain why the product of the reaction above is racemic alanine, rather than the biologically active S enantiomer, L-alanine.

38. Design a synthesis of each of the following target molecules from the starting material given.

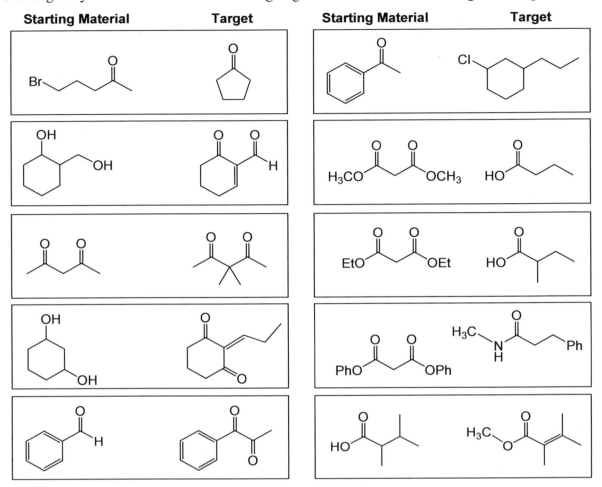

Read the assigned pages in your text, and do the assigned problems.

The Big Picture

The preceding ChemActivities focused on reactions of carbonyl compounds that occur at the carbonyl carbon. This ChemActivity is the first in a series that focus on reactions at the alpha carbon. The common theme in all the alpha carbon ChemActivities is that the alpha carbon is nucleophilic, both in basic conditions (via an enolate) and in acidic conditions (via an enol).

The reactions in the current activity are relatively straightforward (and also less important) than those in the upcoming activities. Before moving on, be sure you are quite comfortable with enolate and enol nucleophiles and skilled at keeping track of your carbons in reactions such as the malonic ester syntheses.

Common Points of Confusion

- Many students assume that the H attached to the carbonyl of an aldehyde is most acidic because it seems like the product (with a negative on a carbonyl carbon) must somehow be resonance-stabilized. In fact, only removal of hydrogens **alpha** to a carbonyl give rise to resonance-stabilized conjugate bases of aldehydes or ketones (called enolates).

- A critical skill going forward (even thought it sounds quite *kindergarten*ish) is to count your carbons after each step in a synthesis or mechanism. It is very easy to lose or gain a carbon during the malonic ester syntheses in this ChemActivity and even easier to make such an error in the aldol condensations in the next ChemActivity. **COUNT YOUR CARBONS** after each step!

Notes

ChemActivity 11: Aldol Reactions

(What is the mechanism of a base-catalyzed **aldol reaction**?)

Model 1: Aldol Reaction

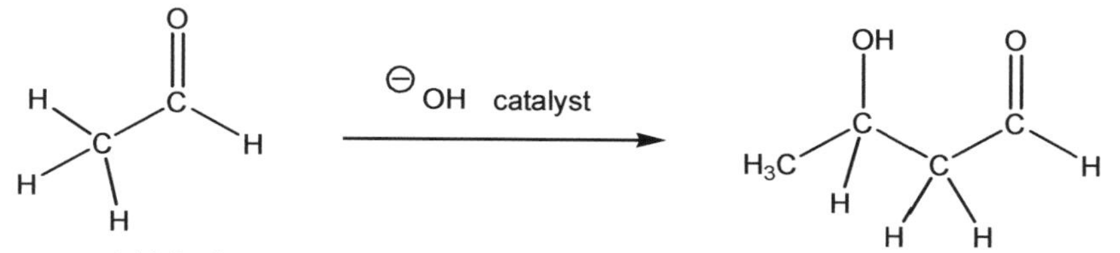

acetaldehyde

an example of an "**aldol**"

Construct Your Understanding Questions (to do in class)

1. Consider the reaction in Model 1.

    a. What two functional groups are found in the product shown in Model 1?

    b. The term aldol is a contraction of "ald" and "ol." Is this consistent with your answer to the previous question? Explain.

    c. (E)How many carbons are there in the aldol product shown in Model 1?

    d. (E)Assuming all carbons in the product above come from the starting material (acetaldehyde), how many molecules of starting material are required to make one molecule of the product?

    e. Label the **new carbon-carbon bond** in the aldol product, and circle each acetaldehyde **subunit** (= two carbons that were originally an acetaldehyde molecule).

    f. Devise a reasonable mechanism for the reaction in Model 1. *Hint*: In the first step, hydroxide reacts with one molecule of aldehyde to produce an enolate nucleophile.

Model 2: Analysis of an Aldol Product Using Retrosynthesis

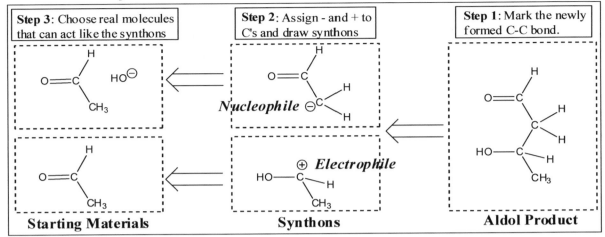

Construct Your Understanding Questions (to do in class)

2. On the aldol product above, complete **Step 1** by drawing a line through the new C—C bond.

3. Explain how a mixture of acetaldehyde and hydroxide is equivalent to the nucleophilic synthon.

4. In what way is acetaldehyde (by itself) similar to (acts like) the electrophilic synthon shown above?

5. Below each, mark the new carbon-carbon bond, and draw a molecule that could be mixed with NaOH to make the aldol product shown.

Careful, this one is tricky!

6. Each of the following aldehydes will **not** produce an aldol reaction when mixed with base. Why?

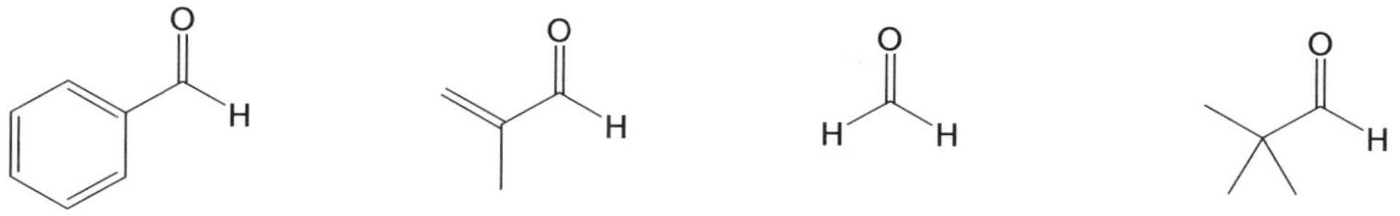

Memorization Task 11.1: An aldehyde must have an H_α to undergo an aldol reaction.

This is because generation of the nucleophile in an aldol reaction involves removal of an H_α.

---

7. (Check your work.) Is your answer to the previous question consistent with Mem. Task 11.1?

Model 3: Mixed Aldol Reactions

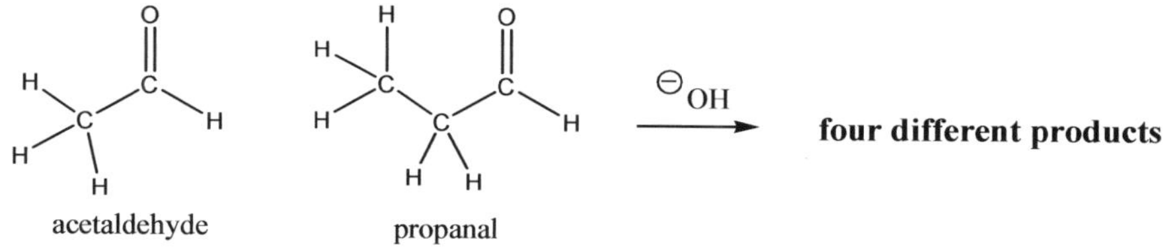

acetaldehyde   propanal

Construct Your Understanding Questions (to do in class)

8. **T** or **F**: In the presence of hydroxide, either acetaldehyde or propanal can serve as a nucleophile in an aldol reaction. Explain your reasoning.

9. **T** or **F**: Either acetaldehyde or propanal can serve as an electrophile in an aldol reaction. Explain your reasoning.

10. Draw the **four products** that result from the mixture in Model 3. *Do not draw the mechanisms.*
(The two products marked "**mixed aldol**" result from a combination of two different aldehydes.)

serves as nucleophile ⇨ serves as electrophile ⇩	(aldehyde structure)	(aldehyde structure)
(aldehyde structure)	Aldol	Mixed Aldol
(aldehyde structure)	Mixed Aldol	Aldol

11. (Check your work.) Do your group's entries above match the answers of groups sitting near you?

Model 4: "Aldol" Reactions Involving Ketones ("Ketols"?)

Ketones can also participate in aldol-type reactions. The product of such a reaction should technically be called a **ketol** (**ket**one + alcoh**ol**), but such products are usually also called aldol products.

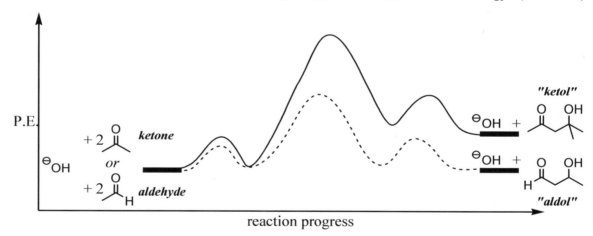

Extend Your Understanding Questions (to do in or out of class)

12. Which is a better electrophile, an aldehyde such as acetaldehyde (Model 1) or a ketone such as acetone (Model 4)?

13. Is your answer to the previous question consistent with the fact that aldol reactions with aldehydes give much better yields than aldol (ketol) reactions with ketones?

14. Energy diagrams for the reactions in Model 1 (---) and Model 4 (—) are shown below. According to these diagrams…

    a. Is the "aldol" reaction in Model 1 **uphill** or **downhill** or **neither** in energy? (circle one)

    b. Is the "ketol" reaction in Model 4 **uphill** or **downhill** or **neither** in energy? (circle one)

15. In a mixed aldol reaction involving acetaldehyde and acetone, only two of the four possible products are observed. Cross out the two products that are unfavorable, and explain your reasoning. (*Hint*: For each product, identify if acetone or acetaldehyde served as the electrophile and nucleophile.)

Memorization Task 11.2: In a mixed aldol reaction involving an aldehyde and a ketone...

the aldehyde <u>or</u> the ketone can act as the nucleophile (assuming an H_α is available to be removed by the base); however, the aldehyde will out-compete the ketone to serve as the <u>electrophile</u> in the reaction.

16. (Check your work.) Is your answer to the previous question consistent with Mem. Task 11.2?

17. For each of the following mixed aldol reaction conditions...

    a.   Write "**possible nucleophile**" below each compound that <u>could serve as the nucleophile</u>.

    b.   Write "**electrophile**" below the compound that is <u>more likely to serve as the electrophile</u>.

    c.   Draw the single <u>most likely</u> aldol product of each reaction.

18. (Check your work.) Are your answers above consistent with the fact that the last two reactions give the *same* major product (3-hydroxybutanal) and that, in both these reactions, the aromatic reactant acts as a spectator and is NOT involved in formation of the major product?

Confirm Your Understanding Questions (to do at home)

19. Give an example <u>not</u> appearing in this activity of an aldehyde that…

    a. <u>Cannot</u> undergo an aldol reaction.

    b. <u>Can</u> undergo an aldol reaction.

20. For each of the following aldehydes, draw the aldol product that will result when the aldehyde is treated with a catalytic amount of base (e.g., KOH).

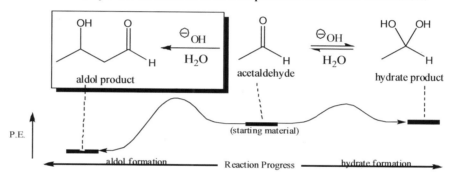

21. The reactants from Model 1 are drawn below. Hydroxide is both a strong base and an excellent nucleophile. As a result, hydrate formation is in competition with aldol formation.

    a. Use curved arrows to show the mechanism of **hydrate** formation.

    b. According to the energy diagram, which is faster, aldol formation or hydrate formation?

    c. Despite the fact that hydrate formation is faster than aldol formation, at equilibrium there is more aldol than hydrate. Explain.

22. Draw the aldol product that results from an intramolecular aldol reaction involving a base catalyst and this dialdehyde.

    a. Explain why, *a dilute solution of a dialdehyde,* favors formation of a cyclic product over a product like the one shown below, right.

    b. If n >3, a *concentrated* solution of a dialdehyde can give the product shown at right, or even longer chains (polymers). Explain this finding.

23. Draw the <u>two</u> different aldol products that result from intramolecular aldol reactions involving the di-aldehyde 3-methylhexanedial and base. For each product indicate which aldehyde of the dialdehyde acted as a nucleophile and which acted as an electrophile. Call the ends the **1-aldehyde** and the **6-aldehyde**.

24. Draw the mechanism of the reaction in Model 4.

25. Forming an aldol from a ketone (also called a ketol) is generally unfavorable. However, in the next activity you will learn a means of pushing such reactions (though we will not isolate the ketol itself). Draw an aldol/ketol for each ketone. For ketones marked with a "2" draw two different products that are constitutional isomers (ignore stereoisomerism).

2

2

assume formation of a four-member ring is unfavorable

26. Draw a ketone not appearing above that could give rise to two different aldol/ketol products, and explain your reasoning.

27. Explain why benzaldehyde cannot act as a nucleophile in an aldol reaction.

benzaldehyde   acetophenone

a. Which will better serve as an **electrophile** in an aldol reaction: **benzaldehyde** <u>or</u> **acetophenone**?

b. Complete the reaction above by drawing the mechanism leading to the **<u>single</u>** most likely aldol or mixed aldol product.

28. Draw a pair of aldehydes and/or ketones (not appearing elsewhere) that would give…

a. a mixture of four different aldol products and draw each product.

b. a high yield of a single aldol product and draw the product.

29. **Use retrosynthesis** to draw aldehyde(s) or ketone(s) that could be used to make each aldol product.

30. Draw the most likely product for each set of reactants or write NO REACTION.

31. The following are called β-keto carbonyl compounds. Construct an explanation for this name.

   a. Draw in the most acidic H or H's (if there is a tie) on each β-keto carbonyl compound.

   b. Draw the conjugate base of each molecule above, including all important resonance structures.

   c. Which is more acidic, the compounds above or an ordinary aldehyde/ketone such as propanal?

   d. Is your answer above consistent with the pK_a data on the table at the end of this book?

   e. In a solution of pentane-2,4-dione, propanal, and base, the mixed aldol product is favored because propanal is a better electrophile, and pentan-2,4-dione is more likely to form a nucleophilic enolate. Draw this product, and show the mechanism of formation.

Read the assigned pages in your text, and do the assigned problems.

The Big Picture

All the chemistry found in this chapter has been introduced elsewhere. An aldol reaction is just a combination of an acid base reaction to form an enolate followed by an addition-elimination at a carbonyl carbon. The fun part of this reaction is that both reactions can take place on an aldehyde and the result is a combining of two molecules of aldehyde. One key challenge is to keep track of your carbons.

Aldol reactions are significant because (unlike reactions involving Grignards, cyanide or other toxic reagents) they provide a way to make and break carbon-carbon bonds in mild conditions. In fact, biological systems use enzyme-catalyzed versions of this reaction for assembling (and disassembling) the carbon backbones of biomolecules. Look for these reactions in any metabolic pathway where the number of carbons changes.

Common Points of Confusion

- Count your carbons!! Get in the habit of counting the number of carbons in the starting material and making sure that all carbons are accounted for in the products. Accidentally losing or gaining a carbon is very easy to do in an aldol reaction.

- The reactions in this chapter involve combining multi-atom molecules. A good strategy is to draw the product as a simple combination of the reactants, even if this means distorting bond lengths and angles. After you have done this you can "clean up" your drawing by redrawing the product with more accurate angles and lengths.

- Retrosynthetic analysis of an aldol product is easier if you first identify the new carbon-carbon bond.

- Some students confuse **retrosynthesis** with a reverse reaction, especially in the case of aldol reactions because aldol reaction are **reversible**. Retrosynthetic analysis of an aldol product is a thought exercise used to figure out which aldehydes or ketones must have been combined to make the target. A reverse aldol reaction is an actual reaction in which the product is hydrolyzed, using water and a catalyst, back into the aldehydes or ketones that were used to make it.

Notes

ChemActivity 12: Aldol Condensations

(Why are some aldol reactions called "condensation" reactions?)

Model 1: (*Review*) Elimination

Construct Your Understanding Questions (to do in class)

1. (*Review*) Use curved arrows to show the mechanism of the reaction in Model 1.

2. (*Review*) What does **LG** stand for, and what is an example of a good LG?

3. (*Review*) Is an E2 reaction **one step** <u>or</u> **two steps** (circle one)?

4. (*Review*) What do the "**E**" and the "**2**" in "E2" stand for?

Model 2: Dehydration (loss of H and OH) in an Aldol Reaction

At higher temperatures (above room temperature) an aldol product undergoes an E2 reaction in base.

Synthetic Transformations 12.1 and 12.2: Aldol Reaction (Base-Catalyzed)

Construct Your Understanding Questions (to do in class)

5. Use curved arrows to show the mechanism of the E2 step in Model 2.

   a. What group (that is usually a poor LG) is serving as a leaving group in this E2 reaction?

   b. Construct an explanation for why this step is often called **dehydration** of the aldol.

c. (Check your work) Is your answer to the previous question consistent with the fact that water is a product of the E2 step?

d. Researchers first studying aldol reactions noticed droplets of water ("condensation") forming on the inside of the flask. Because of this, chemists have come to call aldols, mixed aldols and similar reactions **condensation reactions.** Do you expect these droplets of condensation to form at high temperature, low temperature, or both? Explain.

6. Below is an energy diagram for a ketone aldol (ketol) reaction at high temperature.

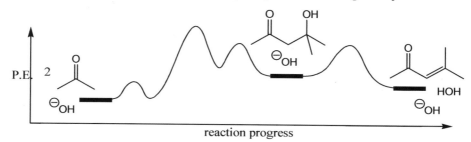

a. Construct an explanation for why each of the following (and the product of the reaction above) is called an α,β-**unsaturated** carbonyl compound. *To help you, the carbons that are "alpha" and "beta" to the C=O are marked on each structure below.*

b. Is formation of the α,β-unsaturated product on the energy diagram overall, **uphill** or **downhill** in energy? [circle one]

c. Construct an explanation for why removing water (boiling it off as steam) will drive this and other condensation reactions to the right to give a very high yield of the α,β-unsaturated product (even though the reaction is slightly uphill in energy).

Memorization Task 12.1: Dehydration step "drives" a condensation reaction to product

Removal of water as steam shifts the equilibrium of an aldol rxn toward the α,β-unsaturated product.

---

7. Consider the α,β-unsaturated product below

Aldehyde　　　　　　　　　　Aldol Product　　　　　　　α,β-unsaturated Product

a. Identify the new carbon-carbon bond that was formed during the aldol reaction (put a line through it), and circle the two aldehyde subunits found in the product.

b. Think backwards using retrosynthesis and draw the aldol product that precedes the α,β-unsaturated product above right. Then think back another step and draw the aldehyde starting material for this reaction.

8. Each of the following was made via an aldol reaction. Mark the new carbon-carbon bond, and draw the carbonyl compound (or compounds) that would give each product (upon heating with base).

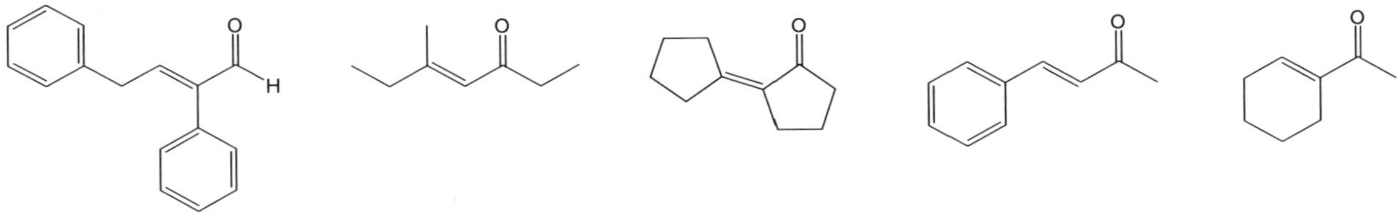

Model 3: Enol as a Nucleophile

Recall from ChemActivity 10 that an enol can function as a nucleophile.
This occurs in acid, where an enolate is unlikely to form.

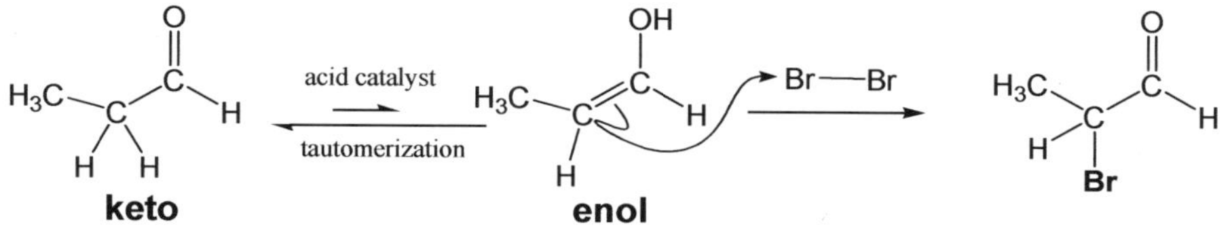

The nucleophilicity of the α carbon of an enol is nicely demonstrated using bouncing curved arrows (as shown above) or using a 2nd-order resonance structure.

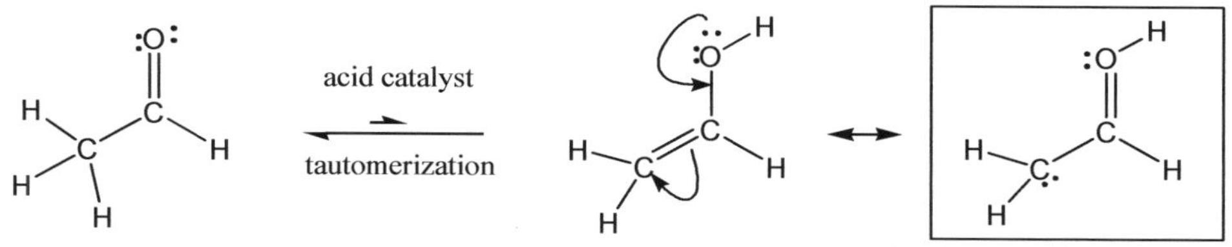

2nd Order Resonance Structure of Enol
(formal charges missing)

Extend Your Understanding Questions (to do in or out of class)

9. Add missing formal charges to the 2nd order resonance structure in the box (lone pairs are shown).
10. Explain how this 2nd order resonance structure emphasizes the nucleophilicity of an enol.

Model 4: Acid-Catalyzed Aldol Condensations

Aldol reactions also work in acidic conditions. However, it is usually **impossible to isolate the aldol product** because, in acid, the reaction tends to go all the way to the α,β-unsaturated carbonyl product.

Synthetic Transformations 12.3: Aldol Reaction (Acid-Catalyzed)

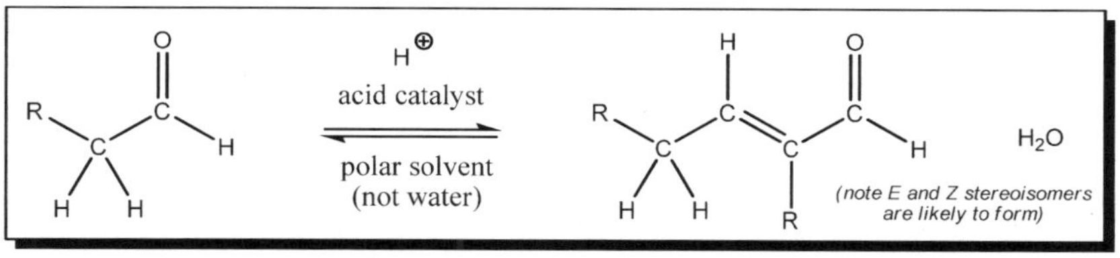

Extend Your Understanding Questions (to do in or out of class)

11. Use curved arrows to show a mechanism for Synthetic Transformation 12.3. *Hint*: the first step is a tautomerization to generate an enol (it may help to draw a second-order resonance structure of this enol).

Model 5: Controlling Product Formation in a Mixed Aldol

Consider the *last* reaction from the previous page…

| acetaldehyde | benzaldehyde | | Product A | Product B |

In the mixture above **Product A** is formed preferentially. This is because **acetaldehyde** <u>can serve as both a better nucleophile *and* a better electrophile</u> compared to **benzaldehyde**.

Extend Your Understanding Questions (to do in or out of class)

12. Using <u>one</u> of the following laboratory procedures you can force benzaldehyde to act as an aldol electrophile and thereby generate **Product B**. Predict which procedure will do this, **(1)** <u>or</u> **(2)**.

(1) A small amount of benzaldehyde is added slowly to a mixture of acetaldehyde & base.

(2) A small amount of acetaldehyde is added slowly to a mixture of benzaldehyde & base.

13. (Check your work.) In one of the boxes above, an aldol reaction (leading to **Product A**) will occur even before addition of the second aldehyde. Cross out this box and explain your reasoning.

14. (Check your work.) In one box above, no aldol reaction can occur *until* a drop of the second aldehyde has been added. Explain.

15. (Check your work) Construct an explanation for why Product A is *unlikely* to form in one of these procedures.

Confirm Your Understanding Questions (to do at home)

16. Draw an example not appearing in this activity of two different aldehydes or ketones that would give a high yield of a single mixed aldol product <u>only</u> when one is slowly added to the other plus base. Specify which carbonyl compound should be added slowly to the other, and draw the product.

17. Construct an explanation for the word **"unsaturated"** in the name "α,β-unsaturated aldehyde" in Model 2.

     a.   What atoms are lost from an aldol in going to the α,β-unsaturated aldehyde?

     b.   Draw the Z stereoisomer that is also formed in the aldol reaction in Model 2.

18. Use retrosynthesis to determine starting materials that will give rise to each product.

(or Z isomer)   (or Z isomer)   (or Z isomer)

19. Draw the product(s) of each of the following acid-catalyzed aldol reactions.

20. Draw all <u>four</u> products that can result when 2-butanone is heated in base at 50°C (RT = 25°C).

21. Explain the following results. When cyclohexanone is treated with base at room temperature a product with molecular formula $C_{12}H_{20}O_2$ is recovered in 22% yield. However, when cyclohexanone is treated with acid at room temperature a product with molecular formula $C_{12}H_{18}O$ is recovered in 92% yield.

22. In some cases the dehydration step of an aldol reaction is very downhill so it is nearly impossible to isolate the aldol product. The reaction between benzaldehyde and acetone is such an example:

     a.   Draw the aldol product and the α,β-unsaturated carbonyl product (both E and Z).

     b.   Construct an explanation for why the α,β-unsaturated carbonyl products are much more stable than the aldol product. (The aldol product cannot be isolated in this case, even if the reaction is run at low temperature.)

23. Even in acid or hot basic conditions a mixture of benzaldehyde and 2,4-dimethylpentan-3-one gives a very low yield of a single aldol product instead of an α,β-unsaturated carbonyl product.

    a. Draw the most likely aldol product that results from this mixed aldol reaction.

    b. Construct an explanation for why, in this case, the α,β-unsaturated carbonyl compound cannot form under any conditions, and even attempting to drive the reaction to the right by boiling off water does not work.

24. One very important thing to remember about the aldol reaction is that it can be **undone**. (It is a reversible reaction.) Draw the mechanism of the reaction below (a reverse aldol).

25. Biological systems use aldol-type condensation reactions to make and break carbon-carbon bonds. For example, plants build fructose (and other six-carbon sugars) from smaller pieces using aldols. Animals (such as ourselves) eat the plants and begin the digestion of fructose by doing a **reverse aldol reaction**, breaking it back into three-carbon pieces.

    a. Fructose is a complicated-looking molecule, but really **it is an aldol in disguise**. Circle the carbonyl and the one OH group on fructose that are positioned like the C=O and OH of an aldol product.

    b. **DIGEST fructose into two pieces!** That is, draw the structures of the two three-carbon pieces into which we break down fructose (and from which plants can make fructose).

26. The preparation of α,β-unsaturated carbonyl compounds is a subject of great interest among synthetic organic chemists. The aldol method is perhaps the most elegant way to generate such molecules, but there are other ways. Design a synthesis of but-2-enal from butanal.

27. Certain compounds rearrange in acidic or basic solution to form α,β-unsaturated carbonyl compounds. Use curved arrows to construct a mechanism for the rearrangement in base (HO⁻), then in acid (H_3O^+) of pent-4-en-2-one into pent-3-en-2-one. (What is the driving force for this?)

28. Design an efficient synthesis of each of the following target molecules. All carbon atoms must come from the starting material(s) given. Be sure to specify the conditions of any aldol reactions.

Target			mix of E and Z	mix of E and Z
Starting material	cyclohexanone formaldehyde		acetone benzaldehyde methyl amine	toluene acetaldehyde

Read the assigned pages in your text, and do the assigned problems.

The Big Picture

The previous activity introduced aldol condensations and mixed aldol condensations. This activity takes these reactions one step farther and explores the E2 reaction that happens at high temperatures or with acid catalysis. When an E2 reaction occurs water droplets form on the inside of the flask. This "condensation" has lead chemists to label these reactions **condensation** reactions.

An aldol reaction may seem new, but it is actually a combination of three reactions that we have already studied. A base catalyzed aldol starts with base removal of an alpha hydrogen followed by nucleophilic addition at a carbonyl carbon, and finishes with an E2 reaction. An acid-catalyzed aldol is the same except that it starts with an acid-catalyzed tautomerization.

At this point in the course you have learned nearly all the basic reaction types. From this point forward, each "new" reaction will actually be a combination of reactions we have already studied. For example, in the next activity we will discover Claisen Condensations which involve all the reactions of an aldol, but add nucleophilic addition-elimination at a carbonyl carbon.

Common Points of Confusion

- In references aldol reactions are lumped together and called condensations (even those which stop prior to the E2 reaction that generates water). At times, the terms **aldol reaction** and **aldol condensation** are used interchangeably.

- Count your carbons!! Get in the habit of counting the number of carbons in the starting material and making sure that all carbons are accounted for in the products. Accidentally losing or gaining a carbon is very easy to do in an aldol condensation.

- This activity is devoted to aldol condensations, and it can be quite easy to deconstruct an aldol product using retrosynthesis *when you are expecting that an aldol reaction has occurred.* Be careful! A simple aldol condensation produces products that look very different from the starting aldehyde, and they can be hard to recognize on an exam when you do not know the question is about aldols.

- The reactions in this chapter involve combining multi-atom molecules. A good strategy is to draw the product as a simple combination of the reactants, even if this means distorting bond lengths and angles. After you have done this you can "clean up" your drawing by redrawing the product with more accurate angles and lengths.

- Retrosynthetic analysis of condensation product is easier if you first identify the new carbon-carbon bond. For dehydrated aldol (condensation) products this is the C=C double bond.

- Some students confuse **retrosynthesis** with a reverse reaction, especially in the case of aldol reactions, because aldol reaction are **reversible**. Retrosynthetic analysis of an aldol product is a thought exercise used to figure out which aldehydes or ketones must be combined to make the target. A reverse aldol reaction is an actual reaction in which the product is hydrolyzed, using water and a catalyst, back into the aldehydes or ketones that were used to make it.

Notes

ChemActivity 13: Claisen & Michael Reactions

(What products result when an ester undergoes an aldol-like reaction?)

Model 1: Claisen Reaction

An aldol-like reaction involving an ester is called a Claisen reaction (or a Claisen condensation) named for German chemist Ludwig Claisen, 1851-1930. The final product is a resonance-stabilized anion.

Synthetic Transformation 13.1: Claisen Reaction

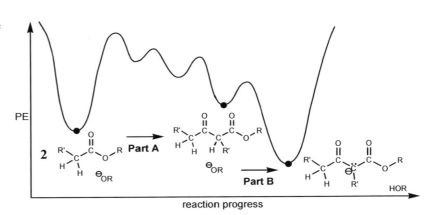

Construct Your Understanding Questions (to do in class)

1. Use curved arrows to devise a mechanism for both parts of the Claisen reaction in Model 1.

2. Draw the other two resonance structures of the Claisen product shown in Model 1.

3. Construct an explanation for why the acid-base reaction in **Part B** is downhill in energy. That is, explain why the Claisen product is lower in potential energy than the intermediate product.

Memorization Task 13.1: An ester needs two H_α's for a Claisen to be favorable (downhill).

---

4. Explain Memorization Task 13.1. (Hint: account for both H_α's on the nucleophilic ester in your mechanism on the previous page.)

Model 2: Choosing the Right Base for a Claisen Reaction

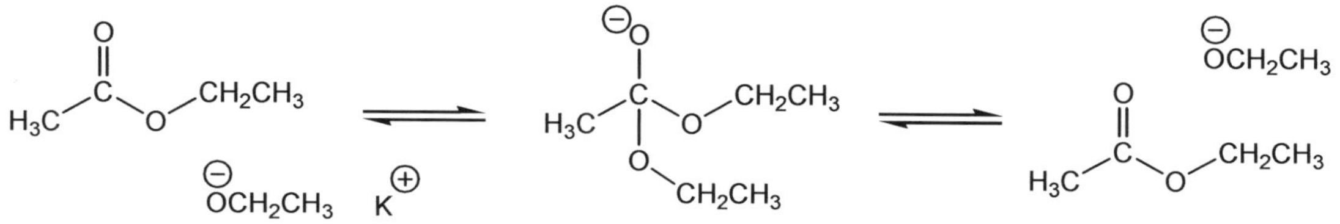

Construct Your Understanding Questions (to do in class)

5. (*Review*) Add curved arrows to the reaction in Model 2.

6. Is ethoxide ion ($CH_3CH_2O^-$) acting as a **nucleophile** or as a **base** [circle one] in Model 2?

7. The reaction in Model 2 takes place at the same time as the Claisen reaction in Model 1, but the former is not very interesting and does not affect the overall outcome of the Claisen reaction. Explain.

8. A student uses potassium ethoxide in the following Claisen Reaction. The result is the unwanted side product shown in the box.

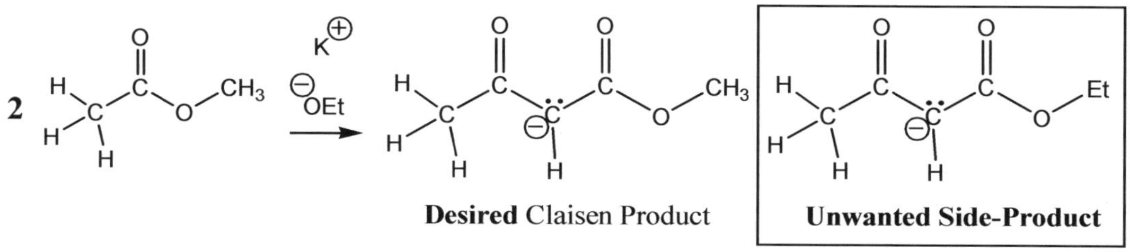

Desired Claisen Product   **Unwanted Side-Product**

a. Describe the key difference between the desired produced and the unwanted product.

b. By what mechanism did this side product come about? (Describe in words. You need not shown the mechanism.)

c.   What base should he have used instead of ethoxide ($^-$OEt) to ensure that only the desired Claisen product was produced? Explain.

Memorization Task 13.2: R group of the base must match R group of the ester in a Claisen

9.   (Check your work) Is your answer to the previous question consistent with Mem. Task 13.2?

Model 3: Protonation or α-Alkylation of a Claisen Product

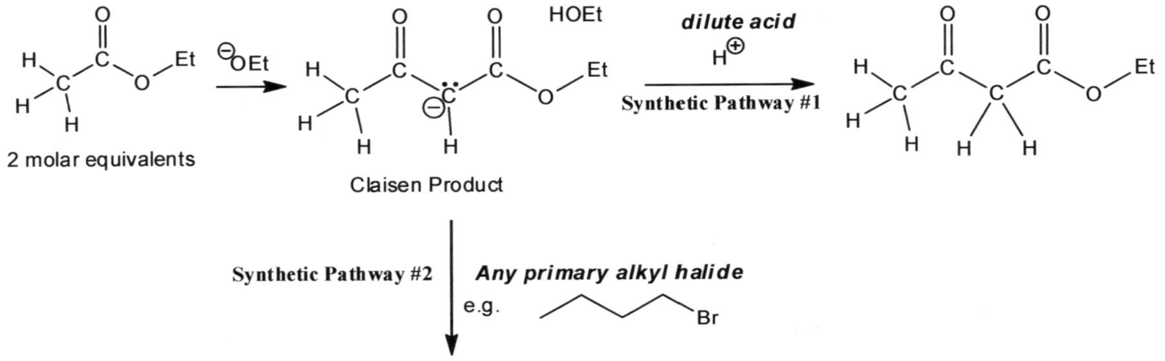

2 molar equivalents

Claisen Product

Synthetic Pathway #1

Synthetic Pathway #2   *Any primary alkyl halide*
e.g.   Br

Construct Your Understanding Questions (to do in class)

10.   Add curved arrows to Model 3 to show the mechanism of Synthetic Pathway #1.

11.   Add curved arrows to show the mechanism of Synthetic Pathway #2, and draw the product.

12.   Construct an explanation for why the products in Model 3 are called **β-keto** esters.

Model 4: Acetoacetic Ester Synthesis of Methyl Ketones

("Bu" = butyl group)

Construct Your Understanding Questions (to do in class)

13. Above each reaction arrow, write the reagent(s) or conditions necessary for that step in the synthesis. (Note: For some steps, not all products are shown.)

14. This ketone can be made using the strategy in Model 4.

    a.   Circle each carbon that was originally part of the starting ester.

    b.   Put a square around each C added during an S_N2 reaction with an alkyl halide (Steps 2 and 4 above), and draw alkyl halides that could have been used to add these carbons.

15. Each molecule of the starting ester in Model 4 (ethyl acetate) has four carbons. Since two molecules of ethyl acetate are used for each molecule of product, this means you begin with <u>eight</u> carbons. Five carbons are lost during the synthesis, so the product contains only <u>three</u> of these eight carbons.

    a.   Check that only three carbons are circled in the previous question.

    b.   Indicate the steps in Model 4 where each of the missing five carbons was lost.

Extend Your Understanding Questions (to do in or out of class)

16. Circle the <u>one</u> ketone below that can be made starting from ethyl acetate using the strategy in Model 4, and briefly explain why each of the other ketones cannot be made using this strategy.

Model 5: Mixed Claisen Reactions

$$\text{(ester structures)} \quad \xrightarrow{\text{KOCH}_3} \quad \begin{array}{c} \text{2 different Claisen} \\ \text{Products} \end{array}$$

Extend Your Understanding Questions (to do in or out of class)

17. Place the labels **can be Nuc⁻** and **can be Elec$^{\delta+}$**, as appropriate, below each ester in Model 5 to show which ester(s) can act as the nucleophile and/or the electrophile in a Claisen reaction.

18. (Above) draw **both** Claisen products that result if the reagents in Model 5 are mixed, and label one of these two products "**mixed Claisen product**" to indicate it was made from two different esters.

19. Suggest a laboratory procedure that gives a high yield of the mixed Claisen product and minimizes the formation of the other Claisen product (the one made from two molecules of methyl propionate).

Model 6: Michael Reaction

Arthur Michael (1853-1942) developed a reaction in which an extra-stable enolate (of a β-dicarbonyl compound) bonds to the β-position of an α,β-unsaturated carbonyl compound. For example...

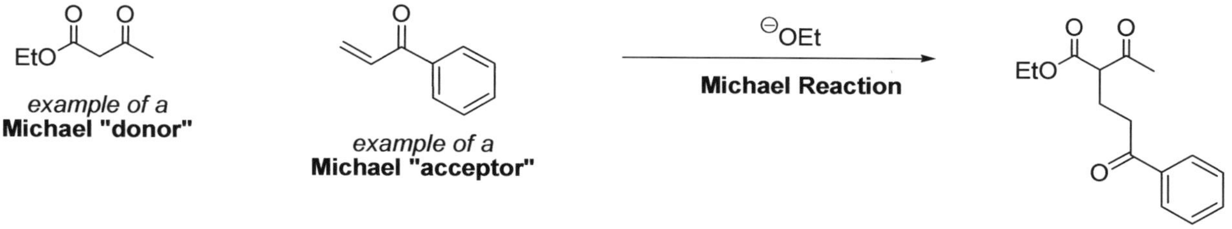

example of a
Michael "donor"

example of a
Michael "acceptor"

Michael Reaction

Extend Your Understanding Questions (to do in or out of class)

20. Mark the new bond in the Michael product, then devise a mechanism for the reaction in Model 6.

Synthetic Transformation 13.2: Michael Reaction

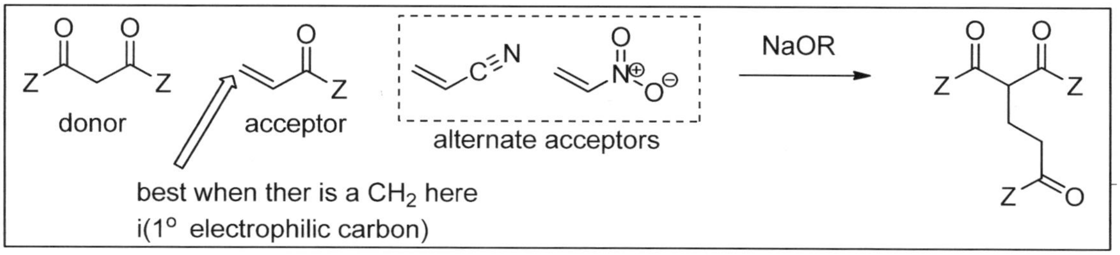

donor

acceptor

alternate acceptors

best when ther is a CH$_2$ here
i(1° electrophilic carbon)

Michael Donors	Michael Acceptors
Can be made from almost any β-keto aldehyde, ketone, ester, nitrile or nitro compound, e.g... 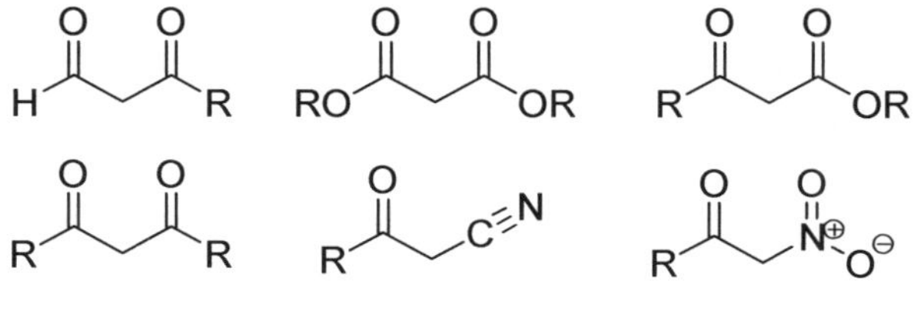	*Almost any α,β-unsaturated carbonyl, nitrile or nitro compound (primary carbon is best), e.g...* 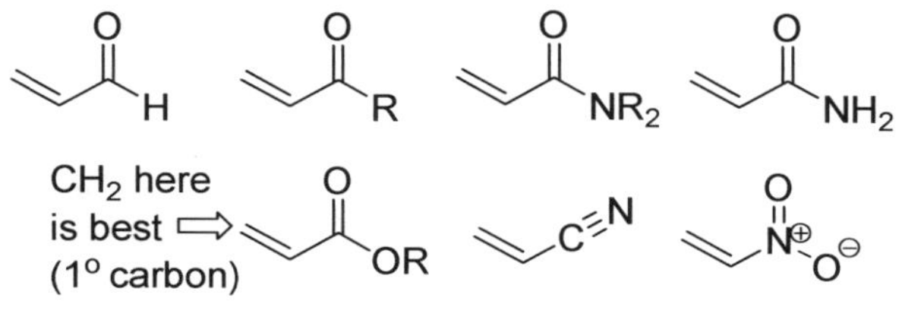

21. Add the label "**nucleophile**" to one of the boxes above to indicate which type of reactant (donor or acceptor) serves as the nucleophile in a Michael reaction. (Label the other box "**electrophile**.")

Confirm Your Understanding Questions (to do at home)

22. Explain why the ester at right does not undergo a favorable (overall downhill) Claisen Reaction. (*Hint*: See Memorization Task 13.4.)

23. What base would you add to each of the following esters to cause a Claisen Reaction?

    a. Show the mechanism of the Claisen Reaction involving the ester on the far right above.

    b. Draw the Claisen product that would result with each ester in the previous question

24. Each set of boxed reactants yields a major product that includes a five- or six-member ring. For each box, draw the best mechanism leading to ring formation.

25. Based on what you learned about EAS-directing effects, which is a stronger **electron-donating group**, an OR group or an R group? (Where R is an alkyl group such as methyl, ethyl, etc.)

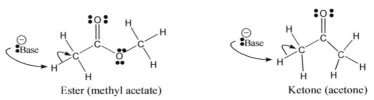

Ester (methyl acetate)   Ketone (acetone)

    a. Which reaction above is more favorable? Explain your reasoning.

    b. Which is more acidic: an ester or a ketone?

    c. Is your answer consistent with the following pK_a data? Ester $pK_a = 24$; Ketone $pK_a = 19$.

Memorization Task 13.3: Aldehydes and ketones are more acidic than esters

Because of this fact, in a mixture containing an aldehyde/ketone and an ester, the aldehyde/ketone is more likely to serve as the nucleophile in a **Claisen-type reaction**.

Note that even though aldehydes and ketones are better electrophiles than esters, the dominant product in such a mixture will usually form via the ester acting as an electrophile. This is because, when an ester acts as an electrophile, a β-dicarbonyl product can form, resulting in the downhill and exothermic loss of the acidic hydrogen alpha to BOTH carbonyl groups. This drives the reaction toward the Claisen product.

---

26. Because aldehydes and ketones are more acidic than esters, in a mixture containing an aldehyde or ketone and an ester, the aldehyde/ketone will act as the nucleophile in a Claisen-type reaction.

    a.   Draw the mechanism of a Claisen-type reaction between acetaldehyde and ethyl formate.

    b.   Draw the aldol product that will also form from this reaction mixture.

    c.   Acetaldehyde is such a good electrophile that it may generate significant amounts of the aldol product. Suggest a laboratory procedure that will nearly eliminate formation of the aldol product from the reaction mixture and maximize the yield of the Claisen product.

27. For each of the following reaction mixtures, list the number of different products that could occur via carbon-carbon bond formation (aldol, Claisen or some mixture).

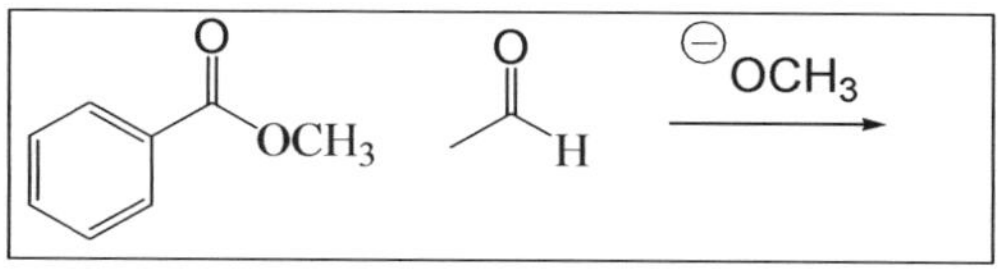

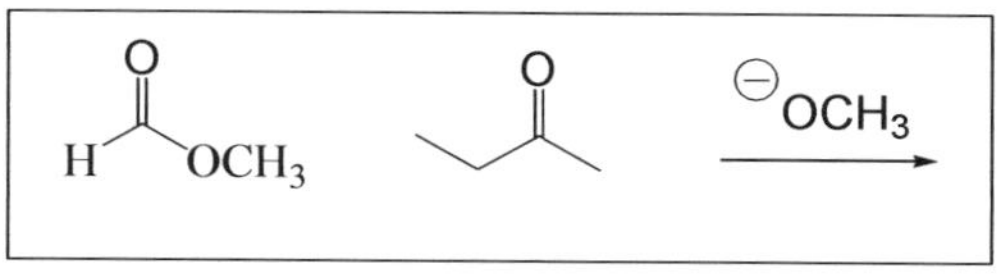

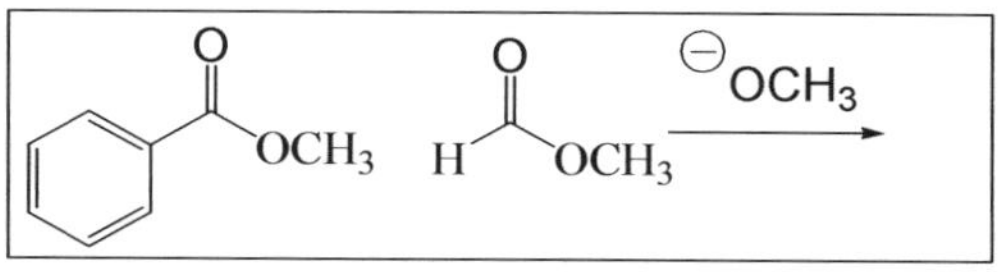

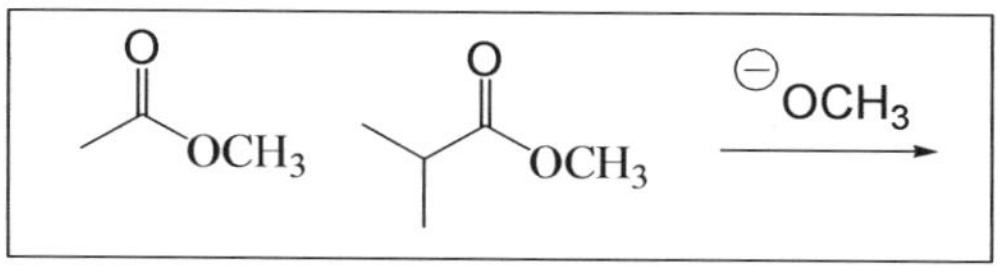

28. Draw the products in the previous question in which an ester acts as the electrophile.

29. Each of the following is the product of an aldol or Claisen-type reaction that has been neutralized with dilute acid. Use retrosynthesis to determine which combination of starting aldehydes, ketones, or esters gave rise to each product, and indicate a suitable base for each reaction.

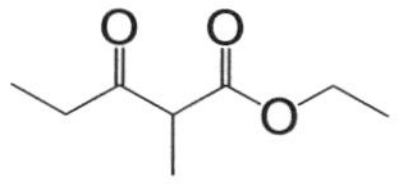

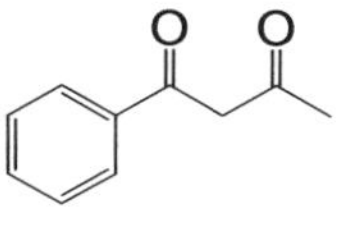

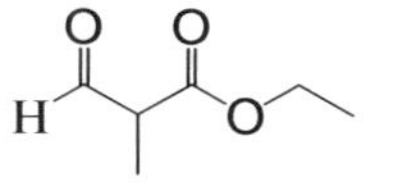

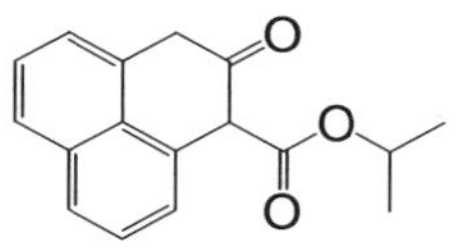

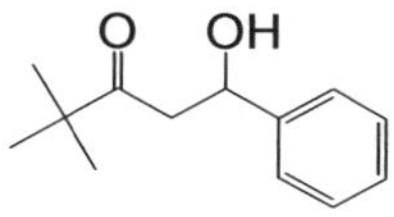

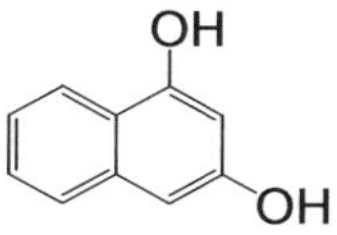

-----(Hint: This molecule adopts a "double enol" form so as to be aromatic. Draw the "double keto" form and consider how this could have been made.)

30. Each target is the product of a Claisen Reaction between two identical esters followed by addition of an alkyl halide. Determine the starting ester and alkyl halide.

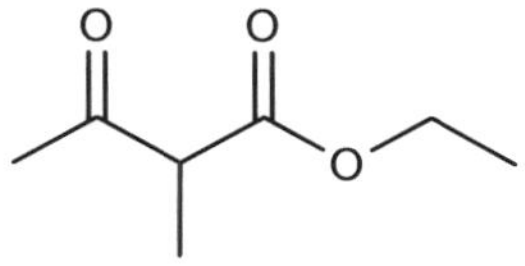

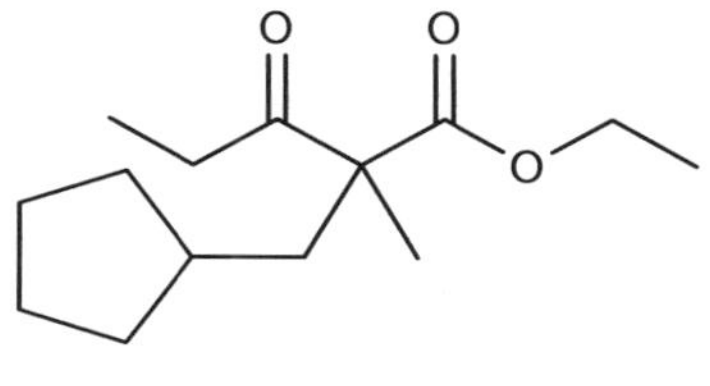

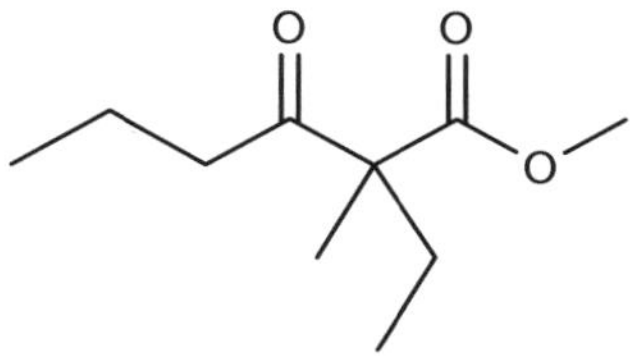

31. Design a synthesis of 3-ethylpentan-2-one using any alkyl halide (with no other functional groups) and any ester (with no other functional groups).

32. Describe key ways in which a Michael addition is like an aldol or Claisen reaction and key ways in which it is different.

33. Draw all reasonable second-order resonance structures for the following α,β-unsaturated carbonyl compound. *Note: Oxygen must always have an octet, even in a second-order resonance structure!*

    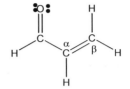

    a. Use these to predict whether the β carbon will be **nucleophilic or electrophilic**.

    b. Is your prediction consistent with how such a compound behaves in a Michael reaction? Explain.

34. Use retrosynthesis to design a way of making each of the following target molecules. The starting materials must include an acetic ester such as ethyl acetate.

Hint: Use 3-methyl-1,5-dibromopentane in your synthesis.	Hint: Use	Hint: The last step is an intramolecular aldol (in acid).
Target 2	Target 3	Target 4

35. Use retrosynthesis to design a way of making each of the following target molecules. The starting materials must include ethyl malonate (shown below).

Ethyl Malonate

OEt = OCH₂CH₃

Target 5

Target 6

Hint: involves a Michael Add'n

36. If you use very high concentrations of 3-methyl-1,5-dibromopentane in the synthesis of Target No. 2, the following side product will also occur. Explain why Target No. 2 is favored over this side product at normal or low concentrations.

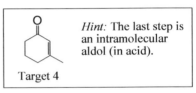

37. Use retrosynthesis to figure out what Michael donor/acceptor pair was used to generate each product. (For the first two, start by marking the new bond in each Michael product.)

Hint: The initial Michael product was treated with H₃O⁺, then decarboxylated to give this product.

38. An **annulation** is a reaction in which a new ring is formed. The **Robinson annulation** is named for English chemist Sir Robert Robinson (1886-1975). It consists of a series of two reactions that use the same catalyst. Under each reaction arrow write the name of a reaction we have studied recently, then devise a mechanism for each step.

Read the assigned pages in your text, and do the assigned problems

The Big Picture

Claisen and Michael reactions will test your ability to keep bonds and atoms straight. All the chemistry found in this chapter has been introduced elsewhere (e.g., enol and enolate nucleophiles, addition-elimination, E2 reactions). The challenge is to apply these to this more complicated environment. One key, as indicated below, is to count your carbons at each step in a synthesis or mechanism.

Aldols, Claisens and related reactions are significant because (unlike reactions involving Grignards, cyanide or other toxic reagents) they provide a way to make and break carbon-carbon bonds in mild conditions. In fact, biological systems use enzyme-catalyzed versions of these reactions for assembling (and disassembling) the carbon backbones of biomolecules. Look for these reactions in any metabolic pathway where the number of carbons changes.

Common Points of Confusion

- Count your carbons!! Get in the habit of counting the number of carbons in the starting material and making sure that all carbons are accounted for in the products. Accidentally losing or gaining a carbon is very easy to do in a Claisen reaction, and especially in a Michael reaction.

- The reactions in this chapter involve combining multi-atom molecules. A good strategy is to draw the product as a simple combination of the reactants, even if this means distorting bond lengths and angles. After you have done this you can "clean up" your drawing by redrawing the product with more accurate angles and lengths.

- Retrosynthetic analysis of a Claisen product is easier if you first identify the new carbon-carbon bond. For a protonated (neutralized) Claisen this is the bond between the alpha carbon and the beta keto group (not the ester carbonyl).

- Some students confuse **retrosynthesis** with a reverse reaction. Like an aldol, a Claisen reaction is **reversible**. Retrosynthetic analysis is a thought exercise used to figure out which esters, aldehydes, or ketones were combined to make the target. A reverse Claisen reaction is an actual reaction in which the product is hydrolyzed back into the starting esters or carbonyl compounds.

Notes

ChemActivity 14: Amines

(How can you make a primary **amine** from ammonia?)

Model 1: pK_a Values for Various Nitrogen Compounds

It is useful to think of base strength as a measure of the <u>likelihood of a lone pair to make a bond to H$^+$</u>.

	acetamide (an amide)	aniline (an aromatic amine)	ammonia (simplest amine)	ethyl amine (a 1° amine)
	O‖ CH₃—C—NH₂	⬡—NH₂	H—N—H ｜ H	CH₃CH₂—NH₂
pK_a of conjugate acid	0	4.6	9.3	10.8

Construct Your Understanding Questions (to do in class)

1. (E)Draw in any lone pairs on the structures in Model 1.

2. According to the pK_a data in Model 1, which nitrogen compound is most basic? …least basic?

3. In general, alkyl groups (such as the Et group on ethyl amine) are electron-donating.

    a. Given this, construct an explanation for why the lone pair on ethyl amine is more likely to make a bond to H$^+$ than the lone pair on ammonia.

    b. Draw <u>di</u>ethyl amine, a secondary (2°) amine, predict whether diethyl amine is a **stronger base** or **weaker base** (circle one) than ethyl amine, and explain your reasoning.

    c. (Check your work.) Is your answer above consistent with the fact that the conjugate acid of diethyl amine has a pK_a of 11.0? Explain.

4. Draw one or more second-order resonance structures that help explain why the lone pair on an aromatic amine (such as aniline) is less likely to form a bond to H$^+$ compared to ordinary amines such as ammonia (NH₃). [See Vol. 2, CA 3 for a definition of second-order resonance structures.]

a. Predict if the pK_a of the conjugate acid of the aromatic compound at right is **closer to 10 or closer to 5**. (Circle one and explain your reasoning.)

"benzyl amine"

b. (Check your work.) Is your answer above consistent with the fact that benzyl amine is NOT considered an **"aromatic amine"** even though it is an aromatic molecule. Explain.

5. Second-order resonance structures are *usually* not as important as ordinary (first-order) resonance structures. In the case of amides, however, there is a second-order resonance structure—shown at right—that is of equal importance to the first-order resonance structure (shown in Model 1). Use this second-order resonance structure to explain why amides are very weak bases.

Model 2: Amine Alkylation

Construct Your Understanding Questions (to do in class)

6. (Review) Use curved arrows to show the mechanism of Rxn 1A and Rxn 1B in Model 2.

7. (E)Label each of the starting amines in Model 2 as 3° (tertiary), 2° (secondary), 1° (primary), or 0° (ammonia).

8. Construct an explanation for why the rate of Rxn 3A > rate of Rxn 2A > rate of Rxn 1A. (*Hint*: See pK_a data in Model 1.)

Memorization Task 14.1: Direct alkylation of an amine is not synthetically useful.

Because a 2° amine is a better nucleophile than a 1° amine, and a 1° amine is a better nucleophile than ammonia, the products of Rxns 1A, 1B and 1C on the previous page will compete with the starting material for reaction with alkyl halide (e.g., methyl bromide). The result is a mixture of all possible products, including the quaternary (4°) ammonium salt shown below. For this reason direct alkylation of an amine is not a synthetically useful option for making primary, secondary or tertiary amines.

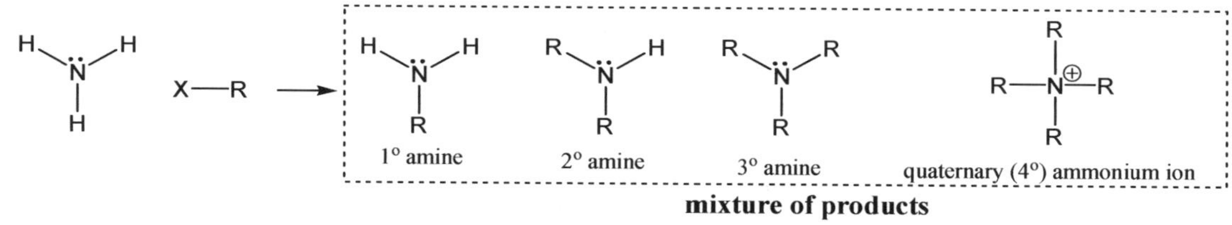

1° amine   2° amine   3° amine   quaternary (4°) ammonium ion

mixture of products

Model 3: Selective Synthesis of 1°, 2°, and 3° Amines

Synthetic Transformation 14.1: Gabriel Synthesis of Primary (1°) Amines

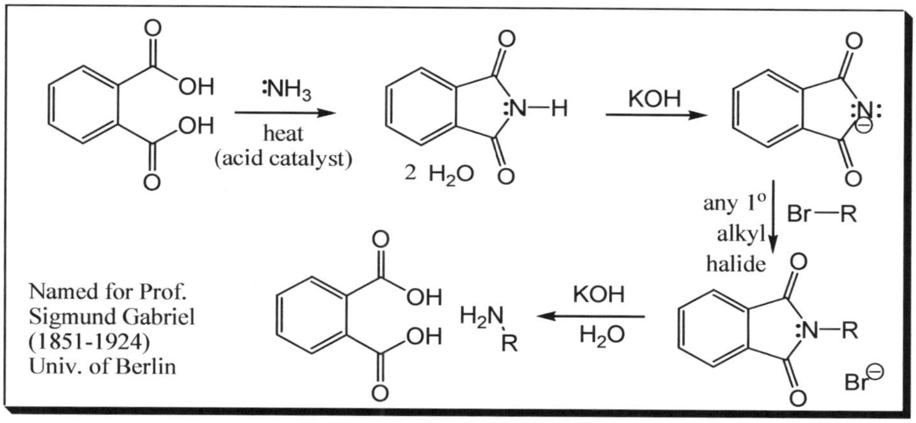

Named for Prof.
Sigmund Gabriel
(1851-1924)
Univ. of Berlin

Synthetic Transformations 14.2-4: Reduction of Amides, Aldehydes, Ketones or Nitriles

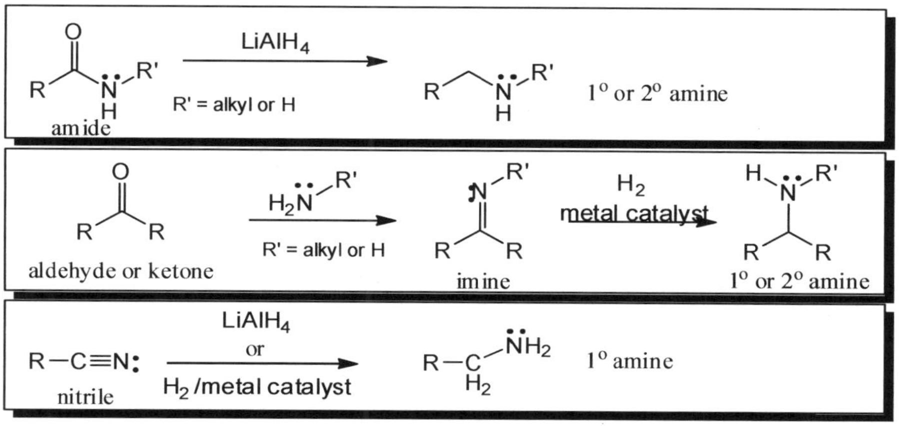

Construct Your Understanding Questions (to do in class)

9.   What do Synthetic Transformations 14.1-14.3 accomplish that the reactions in Model 2 do not?

Extend Your Understanding Questions (to do in or out of class)

10. Show at least one way of making benzyl amine (below) using ammonia (NH_3) as your original nitrogen source. (There are several possible ways, which are covered in the homework.)

11. A conjugate acid-base pair is shown in each box below. For each box, circle the form (conjugate acid or base) you expect to be more abundant in a solution buffered to pH 7 (biological pH).

Model 4: Amino Carboxylic Acids (amino acids)

Twenty common **amino acids** are found in biological systems. Each differs from the others in the identity of the R group. **Proteins** are **polymers** (long chains) of these 20 amino acids. By convention, amino acids (or chains of amino acids called **peptides** or **proteins**) are drawn with the amine end to the left and the carboxylic acid end to the right.

Extend Your Understanding Questions (to do in or out of class)

12. Water solutions are limited to a pH range of about 0-15. Based on the information in Model 4…

   a. Over what pH range will the form of the amino acid shown in Model 4 dominate?

   b. Draw the dominant form of an amino acid at pH 7. Assume that the amino and carboxy ends of an amino acid have the same pK_a values as an isolated amine and an isolated carboxylic acid, respectively.

   c. (Check your work.) In each box on the graph below, draw the form of an amino acid that is expected to dominate over the pH range covered by the box.

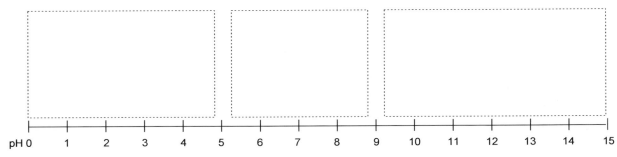

Confirm Your Understanding Questions (to do at home)

13. Draw all three second-order resonance structures of aniline, and explain why aromatic amines are weak bases and poor nucleophiles (as compared to **non-aromatic amines**).

14. Indicate if each molecule is a stronger or weaker base than aniline, and explain your reasoning.

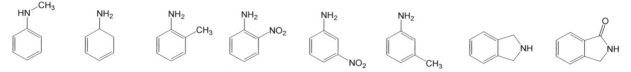

15. Do your best to order the molecules in the previous question from strongest base to weakest base.

16. Use wedge-and-dash bonds to draw a 3-D representation of ammonia.

    a. Label your drawing with the word used to describe the shape of this molecule.

    b. What is the **hybridization state** of the N in ammonia?

17. Ordinary amines <u>cannot</u> be chiral. This is because the lone pair does not hold a fixed position around the N like a bonding pair of electrons. The lone pair of an amine is constantly "flipping" back and forth, changing the configuration of amine.

    a. However, ammonium salts can be chiral. Draw a wedge-and-dash representation of the **R** enantiomer of the conjugate acid of *N*-ethyl-*N*-methylpropylamine.

    b. Consider the following reaction: If a sample of the pure **R** ammonium salt you drew above is treated with base and then acid, will the resulting sample be **pure R, pure S**, <u>or a</u> **racemic mixture**? (Circle one and explain your reasoning.)

18. Place the following labels on the molecules below. Some will have more than one label, some will have no labels. **Aromatic Molecule; Aromatic Amine; Amide**

19. Indicate the approximate p*Ka* of the conjugate acid of each compound above by writing one of…
    pK_a about 0; pK_a about 5; pK_a about 10.

20. Which of the following aromatic compounds is more basic? Explain your reasoning.

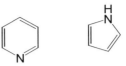

21. Show three different ways of making benzyl amine (see Question 10) using ammonia (NH_3) as your original nitrogen source.

22. In what pH range (if any) is an amino acid likely to...

     a.   have no positive or negative formal charges on any of its atoms?

     b.   be overall neutral?

23. Show the mechanisms of each step in a Gabriel synthesis (See Model 3.).

24. Consider the reaction below.

     a.   Write appropriate reagents over the reaction arrow.

     b.   As a result of this reaction, is the N made more nucleophilic or less nucleophilic?

     c.   Draw second-order resonance structures to support your answer in part b.

25. Biologists often refer to one of the carbons in an amino acid as the alpha carbon. Identify the alpha carbon on the amino acid in Model 4, and devise an explanation for this terminology.

26. Biological amino acids are called L-amino acids. By convention this terminology means that they appear as the amino acid shown in Model 4 (with the H coming out of the page and the R group going into the page when the amino terminus is to the left). However, not all L-amino acids have the same absolute configuration (R vs. S). What is the absolute configuration (R or S) of an amino acid with...

R = CH_3?   R = isopropyl?   R = CH_2OH?   R = $CH_2C_6H_5$?   R = CH_2SH

27. (Check your work.) All of the amino acids above have the same absolute configuration EXCEPT the last one. It turns out that 19 of the 20 naturally occurring L-amino acids have the same absolute configuration. Only cysteine (R = CH_2SH) is different. Explain why cysteine has a different absolute configuration even though, like the others, it has an H coming out of the page and the R group going into the page (when drawn the standard way with the amino terminus on the left).

Read the assigned pages in your text, and do the assigned problems.

The Big Picture

Amines are often the last of the non-biochemical organic chemistry topics. This is because they are a jumping off point for understanding some of the most important biological molecules and their biochemistry. Model 4 begins this exploration with the introduction of amino acids, the building blocks of proteins.

A deep understanding of **biochemistry** requires a deep and mechanistic understanding of organic chemistry. Many students using this book will now go on to study other types of biomolecules. The main ones are sugars (a.k.a. carbohydrates), fats (a.k.a. lipids), and nucleic acids (e.g. DNA, RNA). Those of you who decide to use your foundation in organic chemistry to go onto a biochemistry course will be rewarded for your hard work in this course. Many students who end up in the biological or health sciences say that the reason to learn organic chemistry is to understand the fascinating and dynamic world of biochemistry.

Common Points of Confusion

- pK_a is a central topic in organic chemistry. If you are not comfortable with this topic yet, now is a chance to really hone your understanding. One common point of confusion regarding pK_a's and amines is that the pK_a often associated with an amine is actually the pK_a of the amine's conjugate acid. The only exception to this is if you are talking about taking an H off of a neutral amine. The pK_a associated with this process is 35, indicating that it is very very difficult to remove such an H.

- An amine that is not attached directly to an aromatic ring is called an **aliphatic amine**. The term **aromatic amine** is reserved for a molecule with an amine group attached directly to an aromatic ring. Do not confuse this with aromatic molecules that have an amine group NOT attached to the aromatic ring. Such a molecule (e.g. benzyl amine in Question 4a) might reasonably be called an aromatic molecule with an aliphatic amine.

- Students often forget that, though amines are very good nucleophiles, it is not generally useful to directly alkylate an amine using an alkyl halide electrophile. This is because the product of this reaction (the alkylated amine) is likely a better nucleophile than the starting amine. The result is a mess of products. These difficulties gave rise to the special means of generating amines (especially primary amines) showcased in Model 3.

- Biochemical topics almost always involve stereochemistry. This is because most biological molecules are chiral, and only the correct stereoisomer has the appropriate biological activity. This issue is complicated by the fact that biochemists do not usually use the R/S nomenclature for stereocenters. Instead, the **D/L nomenclature** is used. This simplifies things in certain cases. For example, all 20 of the key amino acids are **L-amino acids**; and this matches up pretty well with the fact that 19 of 20 have an absolute configuration of S. However, since cysteine has a sulfur side chain it causes a priority ranking switch and **cysteine** ends up having an absolute configuration of R. Watch out for this and other disconnects between the D/L naming system and the R/S naming system. (For sugars, the match between the D/L and R/S naming systems is even less close, and is explained in CA 9.)

Notes

ChemActivity 15: Carbon (^{13}C) NMR

BUILD MODELS: 1-bromo-2,3,3-trimethylbutane

(What can a ^{13}C NMR spectrum tell you about the structure of a molecule?)

This activity often requires more than one class period, and can be completed at home or in lab.

Model 1: Spectroscopy Using Radio Waves

In the presence of an external magnetic field, a ^{13}C atom will absorb light in the radio frequency range. The precise frequency of the light absorbed tells us about the atoms near that carbon atom.

For example: the five different carbon atoms labeled **a** to **e** below absorb light of different frequencies. Frequency is measured along the x axis on the chart in a ratio of units called **ppm**. Each cluster of peaks (labeled with a letter) indicates a unique carbon that absorbs light at the frequency indicated on the chart.

When the light is absorbed, it causes what is called a **nuclear spin flip**, so this type of spectroscopy is called **nuclear magnetic resonance** (or **NMR**) spectroscopy.

NMR is the most widely used tool for determining the structure of molecules.

It turns out that the physics of NMR spectroscopy is quite complex. Since its discovery in the 1940s it has grown to become its own subfield of chemistry. The simplified explanation above will be enough for us to learn to interpret NMR spectra.

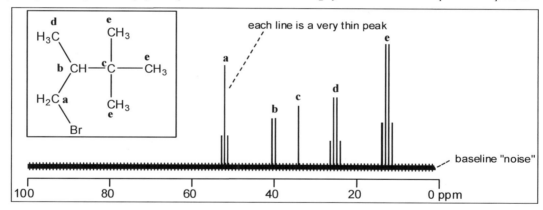

Figure 15.1: Cartoon of (coupled) ^{13}C NMR Spectrum of 1-bromo-2,3,3-trimethylbutane

Construct Your Understanding Questions (to do in class)

1. Next to each carbon in the structure, indicate the number of hydrogens attached to that carbon. Also write this number next to the corresponding **peak cluster** on the spectrum.

2. Construct an explanation for why the three carbon atoms labeled "**e**" are called chemically equivalent (or NMR equivalent) carbons. A model may help.

3. Is there a relationship between the number of hydrogens attached to a given carbon and the number of peaks in the cluster associated with that carbon? If so, what is it?

Memorization Task 15.1: Peak Multiplicity = [Number of H's] + 1

Chemists say that "the two H atoms attached to C_a (we will call them H_a) 'split' the signal of C_a into three peaks." The **multiplicity** (number of peaks in the cluster) of the C_a and C_b peaks is illustrated in the diagram below. The dotted line illustrates what the peaks would look like without any splitting.

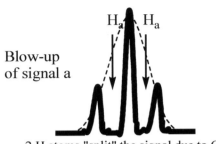

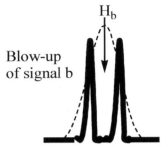

Blow-up of signal a

2 H atoms "split" the signal due to C_a

Blow-up of signal b

1 H atom "splits" the signal due to C_b

A peak cluster with…

- **one** peak is called a **singlet (s)**
- **two** peaks is called a **doublet (d)**
- **three** peaks is called a **triplet (t)**
- **four** peaks is called a **quartet (q)**
- **five or more** peaks is called a **multiplet (m)**

Construct Your Understanding Questions (to do in class)

4. Label each peak cluster in Figure 15.1 with s, d, t, q, or m.

5. Which of the following best explains the placement of peak clusters along the x axis?

    a. The number of hydrogens attached to that carbon atom.

    b. The total number of bonds to that carbon atom.

    c. Distance from the Br atom (Number of bonds away from Br atom).

Memorization Task 15.2: Memorize the following ^{13}C NMR ppm ranges

The property measured by the x axis is called **chemical shift** (think of it as frequency). It is measured in special units called **ppm**.

The chemical shift of an atom is related to the density of the electron cloud around that atom.

Chemical shift is <u>hard to predict precisely</u>. Memorize these **NMR ppm ranges (^{13}C)**.

Chemical shift factors can be synergistic. e.g. an alkene C attached to a halogen (C=C–Br) could be found above 150 ppm.

Type of Carbon	Examples (R = H or alkyl)	ppm Range
C with all single bonds to C's or H's	CH_4 R_3CH	5-60
C with a single bond to O, N, or a halogen	$R_3\mathbf{C}—OH$ $R_3\mathbf{C}—Br$	20-90
C with double/triple bond to C (e.g. C=C)	$R_2\mathbf{C}=\mathbf{C}R_2$	110-150
C with double bond to O (**C=O**)	$R_2\mathbf{C}=O$ RCO_2H	150-220

6.   Consider the following ^{13}C NMR spectrum of 2-bromobutane. Label each peak cluster with a number (1, 2, 3, or 4) indicating its assignment to a specific carbon and a letter (s, d, t, or q).

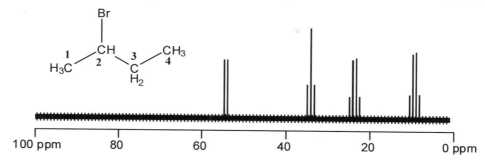

Figure 15.2a: Cartoon of (coupled) ^{13}C NMR Spectrum of 2-bromobutane

Model 2: Coupled versus Decoupled NMR Spectra

If you have a complex molecule with several similar types of carbon atoms, the peak clusters often overlap. This makes interpretation very difficult. To solve this problem, chemists "**decouple**" the H's from the C's. A **decoupled** ^{13}C NMR spectrum of 2-bromobutane is shown below.

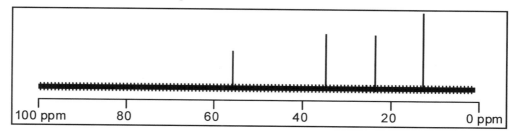

Figure 15.2b: Cartoon of *Decoupled* ^{13}C NMR Spectrum of 2-bromobutane

Think of each C–H bond as a phone link. In the coupled spectrum, each H is "talking to" the nearest carbon, "splitting" the carbon signal through this "phone link." To make a decoupled spectrum, the instrument produces lots of static on the "phone line" so each carbon cannot "hear" the H's attached to it. Consequently, each carbon shows up as a singlet (one peak). You may never see a <u>coupled</u> ^{13}C NMR spectrum outside this activity, though they are presented here as a simple example of splitting in NMR.

Construct Your Understanding Questions (to do in class)

7.   What structural information is lost when we decouple the C's from the H's?

8.   What basic structural information does a decoupled ^{13}C NMR spectrum still convey?

9.   **T** or **F**: In ^{13}C NMR the height of a peak is exactly proportional to the number of equivalent carbons represented by that peak.

10. Sketch a reasonable (coupled) ¹³C NMR spectrum for 3-methyl-1-pentene. Then, below it, sketch the decoupled spectrum. Don't worry about exact placement or relative order of the peak clusters. Just <u>make sure each peak cluster is within the correct range based on Mem. Task 15.2.</u>

11. The NMR spectra in the previous questions are cartoons. Shown below is a real decoupled ¹³C NMR spectrum of 2-bromobutane. Label some differences between the computer-generated decoupled spectrum (Figure 15.2b) and the real version (Figure 15.3, below).

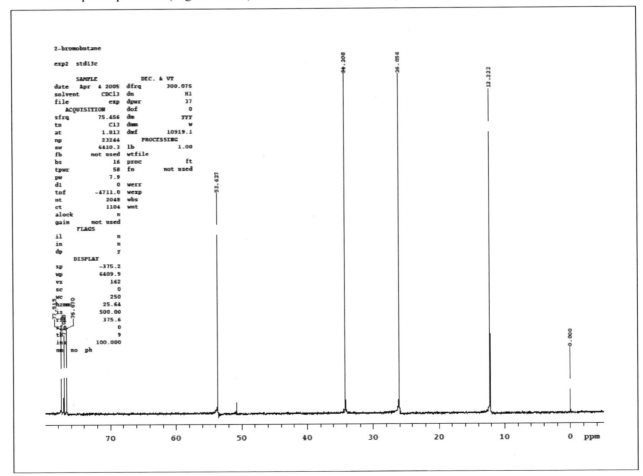

Figure 15.3: Decoupled ¹³C NMR Spectrum of 2-bromobutane

Memorization Task 15.3: NMR Reference and Solvent Peaks

On a ^{13}C NMR spectrum (also called a CMR spectrum) you will often see a peak at 0 ppm and three peaks centered at 77.0 ppm. **These peaks are associated with additives, not the analyte molecule.** The most common additives are **TMS** (a reference compound) and **CDCl$_3$** (a solvent).

TMS (tetramethylsilane)
a reference compound that is often added to the sample because it is inert and gives a peak at 0 ppm that can be used calibrate the ppm scale

CDCl$_3$ (deuterated chloroform)
a solvent that produces three small peaks centered at 77.0 ppm, where few other peaks appear (also can be used to calibrate the ppm scale)

12. Label the peaks in Figure 15.3 due to TMS and CDCl$_3$ with these terms. Also label the impurity peak near 50 ppm that is not associated with the solvent, reference, or the analyte (2-bromobutane).

Model 3: DEPT NMR

Figure 15.4 is a **DEPT** NMR spectrum of 2-bromobutane. A DEPT spectrum consist of **four spectra:** the top spectrum shows only carbons with three H's; the second shows only carbons with two H's; the third shows only carbons with one H; and the bottom shows <u>all carbons with one or more H.</u>

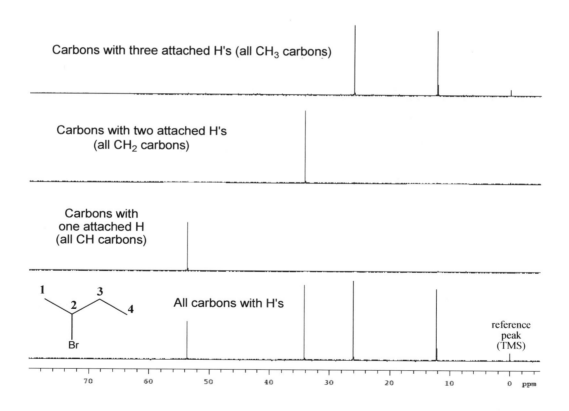

Figure 15.4: DEPT NMR Spectrum of 2-bromobutane

Extend Your Understanding Questions (to do in or out of class)

13. Even though each peak cluster on a DEPT spectrum appears as single peak, you can tell by the level on which it appears if it is a quartet, triplet, doublet, etc. Label each peak with the letter q, t, d, or s, and assign the peaks to carbons 1-4 on the structure of 2-bromobutane on the spectrum.

14. Look at the structure of TMS in Memorization Task 15.3 and explain why the TMS peak appears on the bottom and top levels of the DEPT spectrum.

Memorization Task 15.4: Singlets do not appear on a DEPT Spectrum

By convention, only carbons with H's appear on a DEPT spectrum. To see all the peaks including the singlet peaks you must look at an ordinary ^{13}C NMR spectrum.

---

15. Label each carbon on the structure of 2-butanone with a letter (s, d, t, or q) indicating the type of peak you expect for that carbon

    a. Label each peak on the DEPT spectrum of 2-butanone (below) with a number (1, 2, 3, or 4) assigning it to a specific carbon on the structure.

    b. Which carbon of 2-butanone does not show up on the DEPT spectrum?

    c. In what ppm range on what type of spectrum do you expect to find the peak for carbon 2? (Check your work: See Figure 15.5, *next page*.)

2-butanone

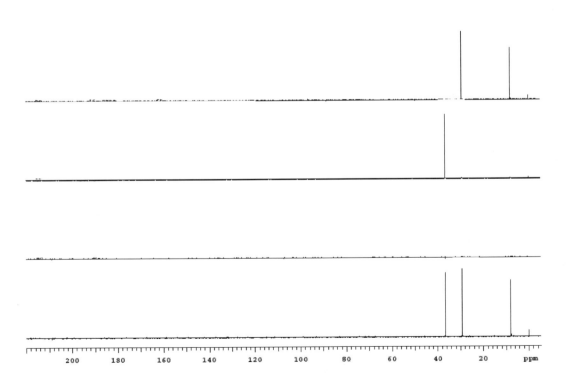

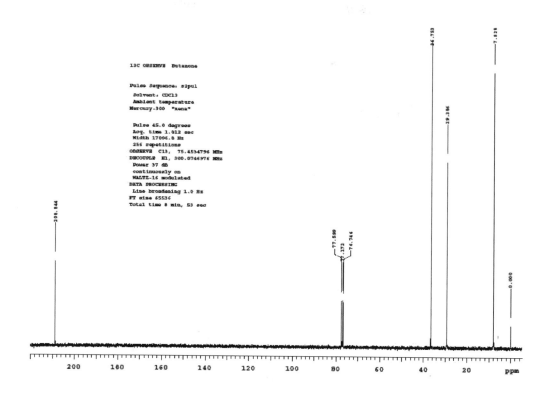

Figure 15.5: Ordinary ¹³C NMR Spectrum of 2-butanone

Model 4: Mirror Planes

Like all two-dimensional objects, Objects 1-3 have a plane of symmetry (also called a **mirror plane**) in the plane of the paper.

For Objects 1-3, some planes of symmetry are shown with dotted lines. Some are not.

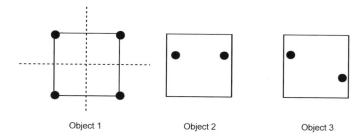

Figure 15.6: Flat Objects Marked with Dots

Extend Your Understanding Questions (to do in or out of class)

16. Mark the planes of symmetry that are NOT marked or noted in Model 4.

   a. Object 1 has two additional planes of symmetry. Mark them.

   b. Object 2 has one additional plane of symmetry? Mark it.

   c. Does Object 3 have any additional planes of symmetry? If so, mark them.

17. Mark each plane of symmetry in the following molecules. For each, indicate <u>if the plane of the paper</u> is a plane of symmetry by writing "POP".

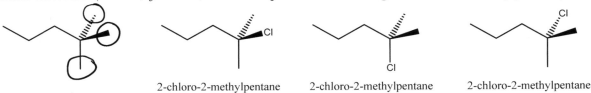

Note: Consider the average of both resonance forms and the average
of single bond rotations when considering symmetry.

Figure 15.7: Molecules with at Least One Symmetry Plane

Model 5: Identifying NMR Equivalent Carbons Using Symmetry

The general test to determine if two atoms (C_A and $C_{A'}$) are NMR equivalent is as follows: draw the structure, but replace C_A with an X. Now do the same replacing $C_{A'}$ with an X. If the two structures are the same *or enantiomers* then C_A and $C_{A'}$ are NMR equivalent.

Example: The three circled carbons are equivalent because replacement of any one with X gives the same molecule. (That is, if X = Cl, all three replacements would give 2-chloro-2-methylpentane)

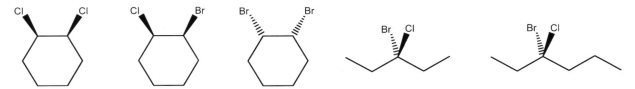

2-chloro-2-methylpentane   2-chloro-2-methylpentane   2-chloro-2-methylpentane

For many molecules, it is easiest to identify equivalent carbons using symmetry: Two carbons are considered NMR equivalent if they can be equated using a mirror plane.

Extend Your Understanding Questions (to do in or out of class)

18. Make a model of 2,2-dimethylpentane (from Model 5), and put it in a conformation so there is a mirror plane equating two of the three circled carbons. Now change the conformation so that a different pair of circled carbons is equated.

19. Number the carbons in each molecule in Figure 15.7 to show which ones are equivalent or distinct. That is, <u>give equivalent carbons the same number</u>. (When you number a molecule to assign its carbons to an NMR spectrum you do not have to follow IUPAC numbering rules.)

20. For the following molecules, confirm that the plane of the paper (POP) is NOT a symmetry plane.

a.   Circle the two molecules that DO NOT contain any symmetry plane. For the others, mark all planes of symmetry using a dotted line.

b.   Number the carbons to indicate which carbons are identical and which are distinct.

Check Your Work: Tips for finding symmetry planes in a molecule

Use the following tips to check your answers to the previous two questions.

- If there are two of something (e.g., Br groups or methyl groups in Figure 15.7), look for a plane of symmetry halfway between them.

- If there is only one of something (e.g., OH group—see Figure 15.7), look for a plane of symmetry that contains that group.

Confirm Your Understanding Questions (to do at home)

21. For each structure:

    a.   Find each mirror plane.

    b.   Number the carbons, giving the same number to equivalent carbons.

    c.   Label each carbon with a letter to indicate the multiplicity of its ^{13}C NMR peak (s, d, t, or q), and where it would appear on a DEPT spectrum.

22. Complete the sentence: In coupled ^{13}C NMR, the number of peaks in a peak cluster ("multiplicity of the peak cluster") tells you…

23. Complete the sentence: In ^{13}C NMR, the location of the peak cluster along the x axis (ppm value) tells you…

24. Complete the sentence: In decoupled ^{13}C NMR, the number of peaks tells you…

25. Complete the sentence: In decoupled ^{13}C NMR, the height of a peak tells you…

26. Chemists rarely use proton-coupled ^{13}C NMR spectra. Explain why.

27. **T or F**: In decoupled ^{13}C NMR each peak cluster is reduced to a singlet (a single peak).

28. The <u>decoupled</u> ^{13}C NMR spectrum of a molecule with the molecular formula C_6H_{12} is shown at right. On the basis of this spectrum, propose a structure for this molecule.

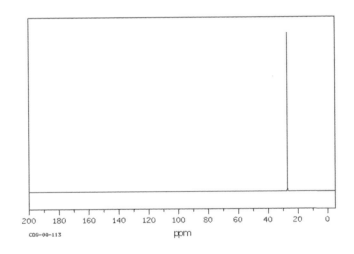

29. For each structure:

- Number the carbons, giving the same number to equivalent carbons.

- Label each carbon with a letter to indicate the multiplicity of its ^{13}C NMR peak (s, d, t, or q), as it would appear on a coupled CMR spectrum.

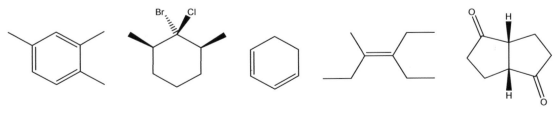

30. Draw the structure of the molecule with molecular formula $C_6H_6O_2$ that is expected to have only the following two carbon NMR peaks: 140.6 ppm, s; and 117.8 ppm, d.

31. Below is the CMR spectrum of a molecule with molecular formula $C_5H_{10}O$. Draw a likely structure of the molecule. The DEPT shows the peaks (from left to right) to be: s, t, q.

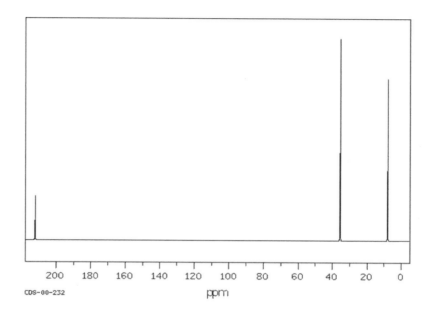

32. The following is a CMR of an isomer of the compound above. Draw a possible structure.

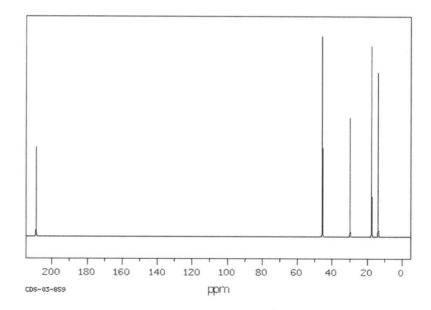

Read the assigned pages in the text, and do the assigned problems.

The Big Picture

NMR (nuclear magnetic resonance) spectroscopy is the most powerful tool that scientists have for looking at the structure of organic molecules. Medicine makes extensive use of this technique–though they drop the word "nuclear" and call it MRI (magnetic resonance imaging). The physics behind NMR is very complex, although it is essentially similar to other spectroscopy such as IR—except that radio frequency light is used instead of infrared light. The very low energy radio waves excite a property called **nuclear spin**. Though many organic chemists do not have a deep understanding of the theory and physics behind NMR, **it is critical that an organic chemist be an expert at interpreting NMR spectra**. Therefore, interpretation of NMR spectra is the focus of this activity and the next.

Interpretation of a (decoupled) carbon NMR spectrum involves two variables: number of peaks, and peak location. The challenge is figuring out how many peaks are expected for a candidate structure; in other words, figuring out which carbons are unique and which carbons are equivalent. The two key tools for this are identifying mirror planes visualizing the molecule in motion.

This activity covers carbon NMR. The next activity covers hydrogen NMR (also called ^{1}H NMR, proton NMR or PMR). In proton NMR, each peak is due to a unique **hydrogen** in the molecule. Organic chemists use the powerful combination of carbon NMR, proton NMR, and MS to deduce the structure of most unknown molecules.

Common Points of Confusion

- **ppm** or **chemical shift**: If a C is close to an electronegative element or involved in a multiple bond, or both, you will find the corresponding peak at higher ppm (farther left on the spectrum). Each chemically distinct C should have a unique chemical shift, though in practice different peak clusters sometimes overlap just by coincidence. This can make interpretation of coupled spectra very difficult. This is why decoupled spectra are usually taken.

- You will not be asked to give the *exact* ppm value associated with a carbon on a given structure. You need only know the ppm ranges. If you are asked to draw a possible structure for a molecule (as in Question 10) just be sure that each peak is within the specified range. You may not even be able to predict the correct order of the peaks within a given range.

- **Peak height** is not a reliable measure of the number of carbons of a given type in C-13 NMR. It is a function of many things, one being whether the C is 1°, 2°, 3°, or 4°.

- **Multiplicity** in a proton-coupled ^{13}C NMR spectrum tells you the number of H's attached to a given carbon. For example, a C that produces a doublet (two peaks) must have one H attached to it. Very often chemists record "proton-decoupled" ^{13}C NMR spectra. Such spectra have a singlet for each chemically unique carbon. This is useful for complex molecules for which the peak clusters would overlap. A decoupled ^{13}C NMR spectrum tells you the number of different carbon atoms present in the sample, but not the multiplicity. This is where a DEPT spectrum can be very useful as it tells you the multiplicity of each peak appearing on the ^{13}C NMR spectrum.

- **DEPT** spectra do not include singlets. This means it is usually necessary to use a DEPT in conjuction with an ordinary ^{13}C NMR spectum. Using the DEPT, you can assign each peak on the ordinary ^{13}C NMR spectum as a siglet, doublet, triplet, or quartet.

- Keep in mind that an NMR spectrum is a "snapshot" of the molecule over a number of seconds. Molecular motion such as rotation of single bonds gets averaged. The second example in Question 17 highlights this. This means the OH group does not break the symmetry of the molecule.

Notes

ChemActivity 16: Proton (^{1}H) NMR

(What can a ^{1}H NMR spectrum tell you about the structure of a molecule?)

This activity often requires more than one class period, and can be completed at home or in lab.

Model 1: Proton NMR (or Hydrogen NMR or ^{1}H NMR)

^{1}H NMR signals are generated by **hydrogen nuclei** and most appear between 0–12 ppm.

[Recall that ^{13}C NMR peaks are generated by carbon-13 nuclei and most appear between 0–220 ppm.]

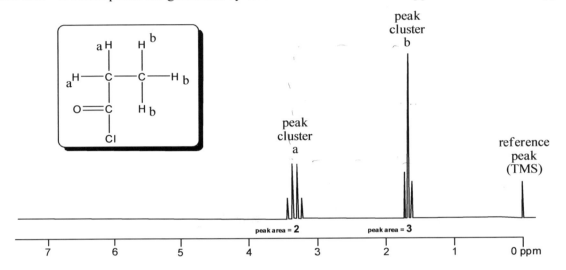

Construct Your Understanding Questions (to do in class)

1. How many peak clusters (excluding the reference peak) are there on the spectrum in Model 1? 2

2. How many different types of chemically distinct H's are there on the structure in Model 1? 2

3. Complete the following sentence: In a proton NMR spectrum there is one peak cluster for each chemically distinct type of (H) or C [circle one].

4. The number listed below a peak cluster gives you the **area (size) of the peak cluster**.

    a. Which peak cluster **a** or (b) [circle one] is bigger (as measured by area)?

    b. What is the ratio of the <u>size of peak cluster a</u> : <u>size of peak cluster b</u>? 2:3

    c. How many H's are there of **type a**? … **type b**? Does this match the ratio above?
       2   3   yes?

    d. Hypothesize: what structural information is conveyed by peak cluster areas in ^{1}H NMR?

ratio of peak clusters tells ratio of Hs associated with those peaks

Memorization Task 16.1: Integration

Peak cluster area, also called the **integral**, **integration**, or **integrated area**, tells you the relative number of hydrogens associated with a given peak, and is often represented by a number written above or below a peak cluster. *Careful! Peak area (not height) tells you the number of H's associated with a peak cluster.*

5. **For now, don't worry about the *number* of peaks in a peak cluster.** We will deal with that on the next page. <u>Based only on the integration of each peak cluster</u> (shown as a bold number below the cluster) assign a letter (**a** or **b**) to each peak cluster, matching it to a type of H on the structure.

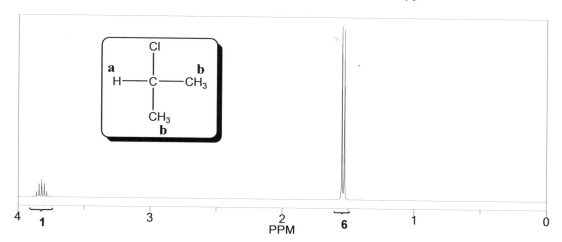

6. Hydrogen atoms attached to the *same* carbon are nearly always equivalent.

    a. Confirm that in each case on this page of two or three hydrogens attached to the same carbon, the same letter assignment is made.

    b. <u>Not all CH₃ (or CH₂) groups are equivalent!</u> Construct an explanation for why the two CH₃ groups on the structure above are equivalent (making all 6 H's labeled "**b**" equivalent), but <u>the two CH₃ groups on the structure at right are NOT equivalent</u>.

Memorization Task 16.2: General Method of Testing if Two Atoms are NMR Equivalent

As with ¹³C NMR, symmetry is an excellent way to identify NMR equivalent H's. The general test to determine if two atoms (H_A and H_A·) are NMR equivalent is also the same as for carbon NMR:

Draw a structure, but replace H_A with an X, then do the same for H_A·. If the two structures are the same *or enantiomers* then H_A and H_A· are NMR equivalent.

Example: The three circled hydrogens are equivalent because replacement of any one with X gives the same molecule. (That is, if X = Cl, any one replacement gives 1-chloro-3-methylpentane)

2-methylpentane   1-chloro-3-methylpentane   1-chloro-3-methylpentane   1-chloro-3-methylpentane

---

7. Draw and name the structure that results if the H with a "?" above it is replaced with a Cl. Based on this, decide if the "? H" is equivalent to a circled H, and explain your reasoning.

8. (Check your work) The "? H" is not identical to the circled H's, but it is identical to five other H's on 2-methylpentane. Identify these five H's that are equivalent to the "? H."

9. **Review Memorization Task 15.1 in the previous ChemActivity**, and answer the following questions about the spectrum from Model 1, reproduced below.

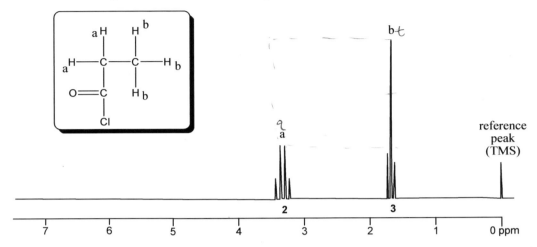

a. Label peak clusters **a** and **b**, above, with s, d, t, q, or m, as appropriate.

b. According to the language in **Mem. Task 15.1 in the previous ChemActivity**, how many times was **peak cluster b** "split/cut" to produce the **three** peaks shown?

twice

10. Just as in ¹³C NMR, splitting in ¹H NMR is caused by nearby H's. Shown at right is a cartoon of peak cluster b from the spectrum above. The dotted line shows the peak as it would be without splitting. Identify the **two** H's on this molecule that are likely splitting this peak into a triplet.

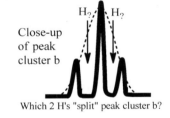

Which 2 H's "split" peak cluster b?

two Ha hydrogens

11. It turns out that in ¹H NMR counting peaks in a cluster tells you the number of neighbor hydrogens within three bonds. (Note equivalent H's do not split each other.)

a. (Check your work) Replace each question mark on the figure above, right with a letter "a" and check that this consistent with your answer to the previous question.

b. Identify the **three** neighbor H's on the structure above which split **peak cluster a** into four peaks.

c. Complete the table at right describing the relationship between peak type and number of neighbor hydrogens within three bonds.

Number of peaks in a peak cluster (peak type)	Number of neighbor H's within 3 bonds
1 (s)	
2 (d)	
3 (t)	
4 (q)	
5 (m)	4 or more

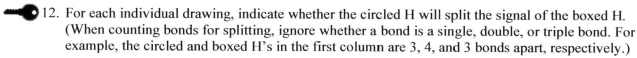

Memorization Task 16.3: Multiplicity in Proton NMR

In ¹H NMR, H's split the signals of foreign (= non-equivalent) hydrogens <u>within three bonds</u>.

This means <u>chemically equivalent H's DO NOT split each other.</u>

12. For each individual drawing, indicate whether the circled H will split the signal of the boxed H. (When counting bonds for splitting, ignore whether a bond is a single, double, or triple bond. For example, the circled and boxed H's in the first column are 3, 4, and 3 bonds apart, respectively.)

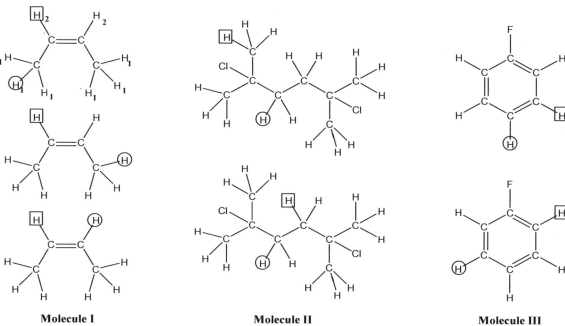

| | **Molecule I** | **Molecule II** | **Molecule III** |

13. On the first row of structures above…

   a. Label each hydrogen on the structure with a number, giving identical hydrogens the same number. (The first one is done for you.)

   b. For molecules II and III, complete a table, like the one for Molecule I, below, by reporting the **integration** and **multiplicity** of the peak due to each type of hydrogen.

Molecule I	Integration	Multiplicity
H₁	6*	d
H₂	2*	q

*Ratio of H₁ to H₂ could also be reported as 3:1

14. (Check your work) Explain why, for Molecule II, the H's on carbons 3 and 4 do not split one another even though they are only three bonds apart.

15. (Check your work) There are no CH₂ (methylene) groups on Molecule III, yet the signal due to the H on carbon 4 is a triplet. <u>Explain how a triplet can occur with no CH₂ group next door.</u>

16. On a proton NMR spectrum, the chemical shift (ppm/placement along the x axis) conveys similar information as in ^{13}C NMR. Based on this, construct an explanation for why **peak cluster a** is farther to the left (higher ppm) than **peak cluster b** on the spectrum in Model 1.

Memorization Task 16.4: Memorize Key ¹H NMR Chemical Shifts (ppm ranges)

Type of Hydrogen	Examples (Marked H's are within the given NMR ppm ranges)	ppm Range
H three or more bonds from a functional group		1 - 2
H two bonds from π bond (C=C or C=O)	benzylic / allylic / "alpha" to a carbonyl	2 – 3 (1.8-2.6)
H two bonds from O, N, or halogen		3 – 4 (2.5-4.5)
H attached to C=C (not benzene)		4 – 7 (4.5-7.0)
H attached to benzene ring		7 – 9 (6.5-8.5)
H attached to carbon of C=O (carbonyl)		10 (9.7-10)
H attached to O or N	water / alcohol / amine / phenol (Ar—OH)	1 - 8
	carboxylic acid	10 - 13

Extend Your Understanding Questions (to do in or out of class)

17. Alcohols do not always follow the rules for predicting multiplicities. <u>Based on integration and chemical shift alone</u>, assign each of the three (non-reference) peaks in the proton NMR spectrum of ethanol at right.

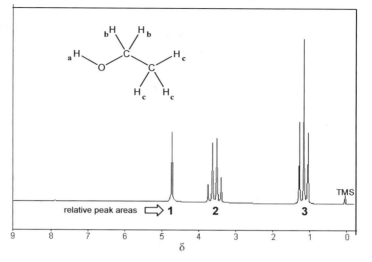

18. Write the expected multiplicity above each peak cluster associated with ethanol. Which clusters have a multiplicity that does not fit the rules we have learned so far?

Memorization Task 16.5: Special Considerations for H Attached to N or O

- Under normal conditions, the H of an alcohol or amine does not participate in splitting.

- When D_2O is added to a NMR sample, a peak due to an H attached to O or N will disappear.

---

19. Does Memorization Task 16.5 explain the inconsistencies you noted above? Explain.

20. In a typical proton NMR sample, the solvent is present in much higher concentrations than the molecule being analyzed. Construct an explanation for why this means that any solvent with a hydrogen (such as $CHCl_3$, C_6H_6, CH_2O, etc.) is *not* useful as a proton NMR solvent.

21. $CDCl_3$ (deuterated chloroform, or chloroform-d) is the most common NMR solvent because it is cheap relative to other deuterated solvents, dissolves many organic molecules, and has no H's. Most $CDCl_3$ is contaminated with a very small amount of $CHCl_3$. (The H of $CHCl_3$ appears as a singlet at 7.24 ppm.) Explain how this peak removes the need to add TMS to the sample.

22. Cross out the <u>two</u> molecules below that cannot serve as a proton NMR solvent.

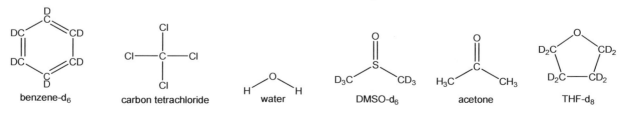

Confirm Your Understanding Questions (to do at home)

23. Identify/assign each peak (including any solvent or reference peaks) on the following spectrum.

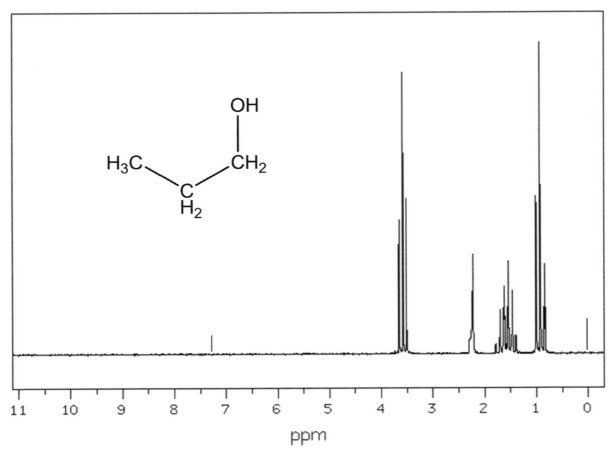

24. For the spectrum above, does the alcohol H participate in splitting (as described in Memorization Task 16.5)?

25. (Check your work) Are your assignments above consistent with the fact that only the peak at 2.23 ppm disappears when the sample is treated with D_2O? Explain.

26. Assign each peak cluster on the following spectrum, and predict the integration value for each.

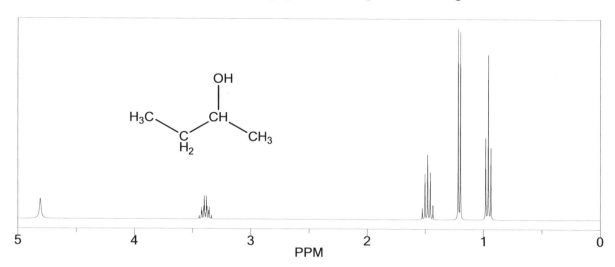

27. The proton NMR data for 1-bromopropane are as follows: H_a: triplet (2H) 3.32ppm; H_b: multiplet (2H) 1.81ppm; H_c: triplet (3H) 0.93ppm. (Relative integrations shown in parentheses.)

   a. Through how many bonds can a hydrogen split another hydrogen?

   b. According to this splitting rule, does H_a split H_c?

   c. Is your answer in part a) consistent with the multiplicity listed for peak clusters a and c?

   d. How many hydrogens split H_b?

   e. Upon **very close** inspection of the proton NMR spectrum of 1-bromopropane, you would find that peak cluster b has at least six peaks. Is this consistent with your answer in part d)?

   f. Speculate as to why any peak cluster with more than four peaks is listed simply as a "multiplet."

28. For each structure below, use letters or numbers to indicate chemically equivalent and distinct hydrogens, and make a table showing the predicted integration and multiplicity of each peak cluster.

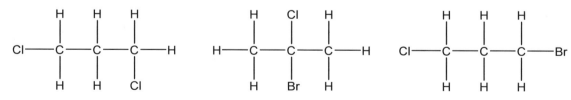

29. For each structure below, use numbers to indicate chemically equivalent and distinct hydrogens, and make a table showing the predicted integration and multiplicity of each peak cluster.

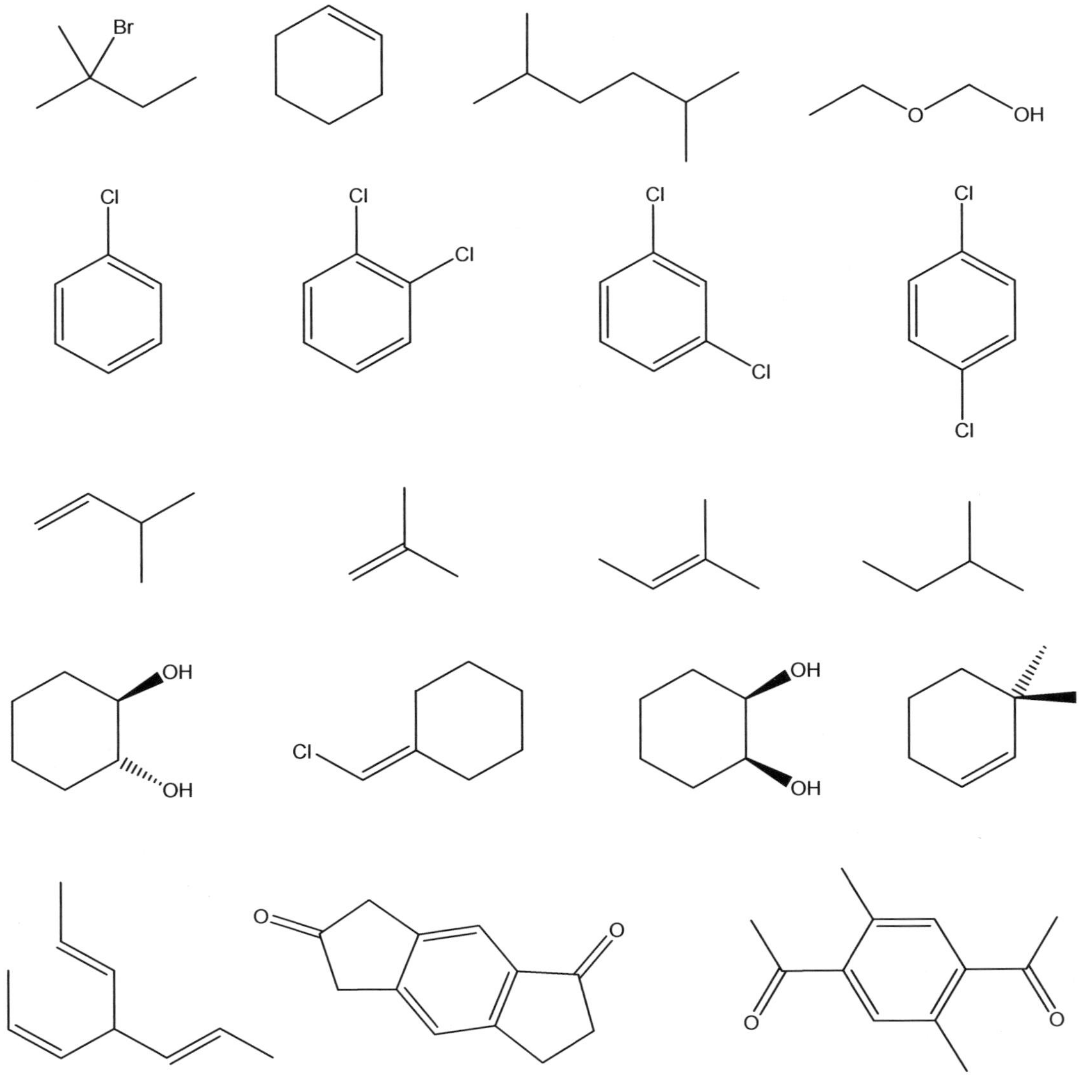

30. Imagine you have two bottles: one with (R) and the other with (S)-2-bromobutane. Unfortunately, your lab partner messed up and labeled both bottles simply "2-bromobutane." Can you use NMR to sort out this problem? Explain why or why not.

31. A researcher wants to take a proton NMR of a molecule that dissolves only in water. What solvent should she use to make the NMR sample?

32. Based on the ppm range in which the peaks appear on the spectrum below, what can you infer about the molecule associated with the spectrum?

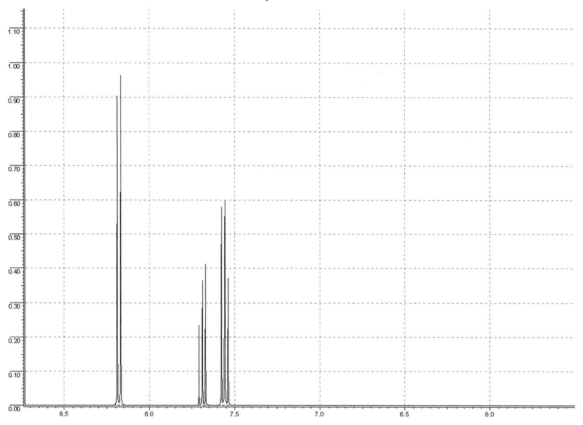

33. Propose a structure to go with the spectrum above assuming the unknown has molecular formula $C_6H_5NO_2$, and the integrations of the peaks are 2:1:2, from left to right.

34. Propose a structure to go with the spectrum below assuming the unknown has molecular formula C_4H_8O, and the integrations are as shown. (*Hint*: First determine the degrees of unsaturation.)

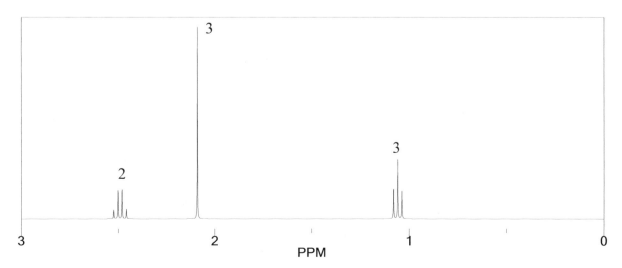

35. (Check your work) The peaks at 1.1 and 2.5 ppm on the previous page represent the <u>signature pattern of an ethyl group</u>. This pattern (a quartet worth 2H's and a triplet worth 3 H's) is so common it is worth memorizing.

    a. Find this pattern on the proton NMR spectrum of ethanol in CTQ 20.

    b. Sketch the proton NMR spectrum expected for diethyl ether ($CH_3CH_2OCH_2CH_3$).

36. Propose a structure to go with the proton NMR spectrum and the following mass spectral data: $[M]^+$ = 136 (100), $[M+1]^+$ = 137 (3.3), $[M+2]^+$ = 138 (97.3)

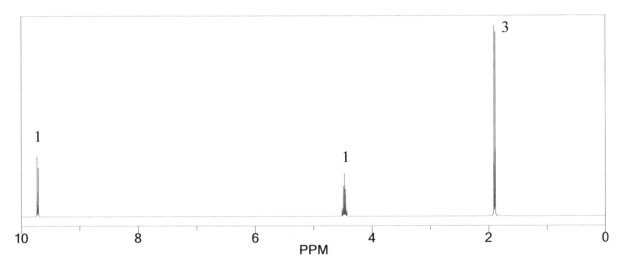

37. Construct an explanation for why the peak in the spectrum above found at 4.5 ppm is outside the typical range for a hydrogen alpha to a carbonyl (2-3) *and* the typical range for a hydrogen bound to the same carbon as a halogen (3-4).

38. Draw the structure of an unknown to go with this proton NMR and the following mass spectral data: $[M]^+$ = 166 (50), $[M+1]^+$ = 167 (5.5), $[M+2]^+$ = 168 (16.1)

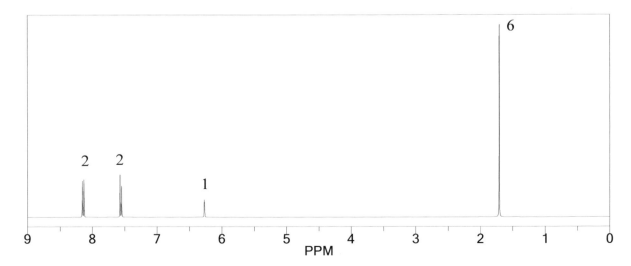

39. Peak clusters in the 7-9 ppm region are almost always indicative of H's attached to a benzene ring. (As we will learn later, the region from 7-9 ppm is called the aromatic region, and molecules containing a benzene ring are by far the most common type of aromatic molecule).

Peak clusters in the aromatic region often overlap one another. For example, the mass of peaks between 7.3 and 7.4 ppm on the spectrum below is actually two overlapping peak clusters: a doublet of doublets worth 2H's, overlapping with a triplet worth 1H.

Draw one structure that goes with both spectra on this page.

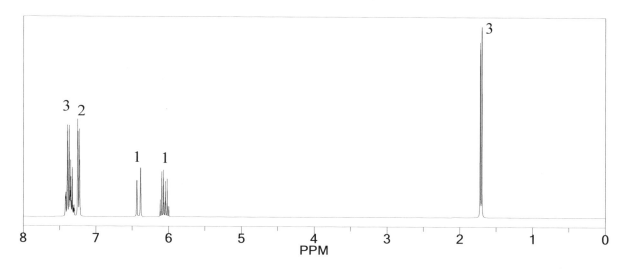

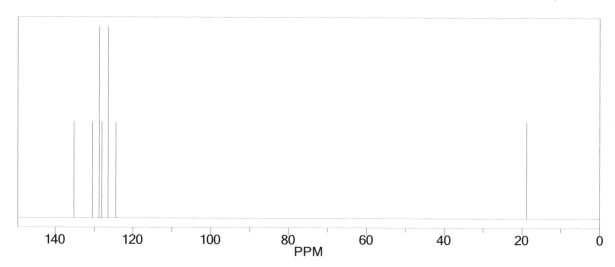

40. (Check your work) How can you tell the top spectrum on this page is a proton NMR spectrum and the bottom spectrum is a carbon NMR spectrum?

41. (Check your work) According to the CMR spectrum, how many unique C's are there?

42. (Check your work) Explain why, on the proton NMR spectrum, the peak at 6.1 is expected to be a doublet of quartets.

43. The spectrum below goes with one of the diasteriomers shown. The other diasteriomer is the answer to the previous question.

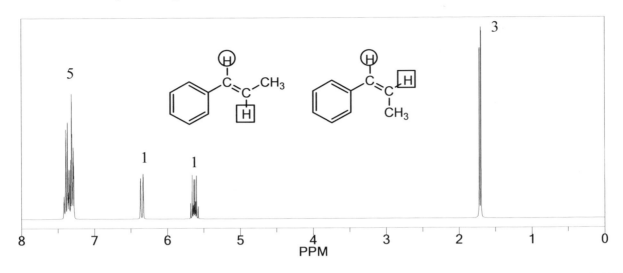

Memorization Task 16.6: Splitting between *trans* H's is larger than between *cis* H's

For reasons we will not discuss here, splitting between two H's that are *trans* to each other is larger than splitting between two H's that are *cis* to each other. This is often used to identify whether a molecule is *E* or *Z*.

---

a. Splitting between the circled and boxed H's below can be measured by looking at either the peak cluster of the circled H or the boxed H; however, the cluster associated with the boxed H is much harder to interpret since it is split both by the circled H and the methyl group. Identify the peak cluster that goes with the circled H, and confirm that ONLY the boxed H is near enough to split this peak.

b. On which proton NMR spectrum (the one above or the one on the previous page) is the splitting between the circled and boxed H's largest?

c. Cross out the structure above that does NOT go with the spectrum, and explain your reasoning.

d. Number the H's on the remaining structure and use these to assign each unique type of H to a cluster on the spectrum. (Note that there are some overlapping clusters near 7.4 ppm.)

e. (Check your work) The structure that you crossed out above goes with the NMR spectra on the previous page. Draw this structure on the previous page, and assign each H to a peak cluster on the proton NMR spectrum and each C to a peak on the carbon NMR spectrum.

Read the assigned pages in the text and do the assigned problems.

The Big Picture

Proton NMR works in much the same way as carbon NMR, but again interpretation of the spectra is a much more important skill for an organic chemist than understanding the complex physics behind the instrument. The key elements of a proton NMR spectrum are reviewed below.

Common Points of Confusion

- Equivalent H's do NOT split each other. The rule itself is not hard to follow, but students sometimes forget to check for symmetry and so do not realize that two neighboring H's are equivalent. (*See* Molecule II in Question 12)

- Peak area, not peak height, tells you the relative number of H's associated with a peak cluster. This leaves open the possibility that a tall skinny peak could be smaller (in terms of area) than a short fat peak.

- Splitting in carbon NMR tells you the number of H's attached to a given carbon. For this reason, students incorrectly assume that splitting in proton NMR tells you the number of H's associated with a peak cluster. In fact, it is a bit more complicated (see last bullet).

The following is a summary of the key elements of proton NMR:

- **ppm** or **chemical shift** (*given by location along the x axis*) is a function of the amount of electron density around an H. The closer the H is to an electronegative element, the more "**deshielded**" it is and therefore the higher the ppm number of its peak cluster (farther left on the spectrum). Multiple bonds also cause the signal of nearby H's to be shifted to the left. Memorization Task 16.4 gives chemical shifts for common functional groups. Each chemically distinct H is expected to have a unique chemical shift, though in practice different peak clusters sometimes overlap just by coincidence. This is a bigger problem in proton NMR than in carbon NMR (especially in the so-called "aromatic region" from 7-9 ppm), since most proton NMR peaks are squeezed into just an 8 ppm range (1-9 ppm).

- **Integration** or **peak cluster area** (*given by a number above or below a peak cluster, or by a line stepping up from left to right called the integration line*) tells you the relative area of each peak and therefore the relative number of equivalent H's represented by each peak. Note that integration (especially the integration line) gives you only a ratio of peak areas. This means the same integration may be reported, for example, on the spectrum of a molecule with 1H to 3H ratio and a 2H to 6H ratio.

- **Multiplicity** is the number of peaks in a peak cluster (also called **splitting** or **proton-proton coupling**). It tells you the number of nonequivalent neighbor H's within three bonds. For example, a doublet (two peaks) tells you there is exactly one non-equivalent H within three bonds of the H responsible for this signal.

Notes

Nomenclature Worksheet 3

NAMING BENZENE DERIVATIVES

Model 1: Naming Mono-Substituted Benzene Rings

Simple mono-substituted benzene rings can be named by placing the substituent name in front of benzene →

chlorobenzene

nitrobenzene

ethylbenzene

ethoxybenzene

Memorization Task NW3.1: Parent Names of Common Benzene Derivatives

A handful of these are so common that they constitute their own parent name. Know the following…

toluene

phenol

aniline

benzaldehyde

benzoic acid

When a substituent contains an alcohol (OH), amine (NH_2), or in some cases a carbonyl, the substituent is considered the parent chain, and the ring is considered an attached group called a **phenyl group**.

For other functional groups (e.g., alkenes, halogens, etc.) the benzene ring is often the parent chain.

3-methyl-1-phenylbutan-2-ol

4-methyl-5-phenylhexan-3-amine

(*E*)-pent-3-en-2-ylbenzene

(3-bromo-2-methylpentyl)benzene

Questions (to do in or out of class)

1. What is the name most often used for aminobenzene?

2. Name each of the following compounds.

3. Draw each of the following compounds from their names.

   a. 4-phenylbutan-2-ol   b. pent-3-yn-2-ylbenzene   c. (E)-4-phenylbut-3-en-2-ol

Model 2: Naming Multi-Substituted Benzene Rings

Either the *ortho/meta/para* system or the numbering system may be used for naming benzene rings with two substituents. When there are three or more substituents the numbering system must be used.

3-methyl-1-nitrobenzene
or
meta-nitrotoluene
or
3-nitrotoluene

2-aminophenol
or
ortho-aminophenol
or
ortho-hydroxyaniline, etc.

2,4-dimethylbenzaldehyde

2-hydroxy-4,6-dimethylbenzoic acid

Note that *ortho, meta,* and *para* are often abbreviated *o, m,* and *p,* respectively.

Questions (to do in or out of class)

4.   Write another name (not appearing in Model 2) for 2-aminophenol.

5.   Draw the structure that corresponds to the name 2-phenyl-6-methylphenol.

6.   Explain why the following is not a correct name, 4,5-dimethylphenol.

7.   Name each of the following compounds.

8.   Draw each of the following compounds from their names.

   a. *p*-benzenediol   b. 2,4,6-trinitrotoluene   c. (E)-*ortho*-(prop-1-enyl)phenol

Confirm Your Understanding Qustions (to do at home)

9. Name each of the following compounds.

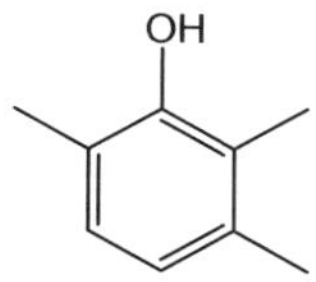

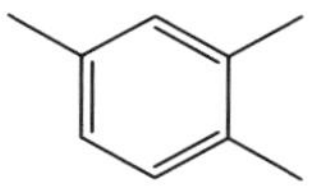

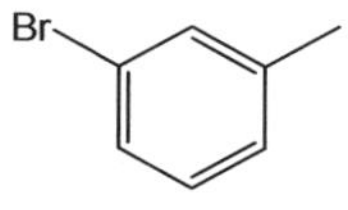

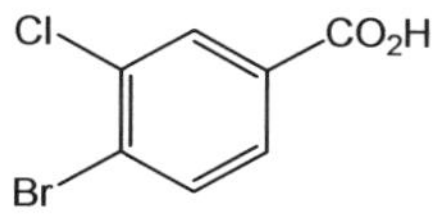

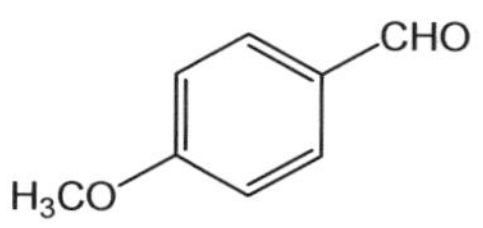

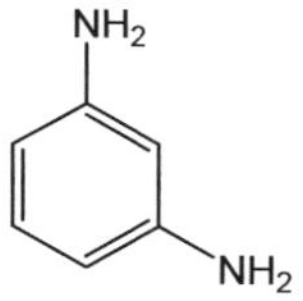

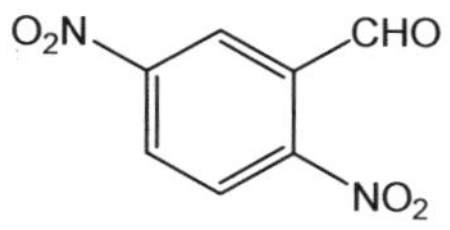

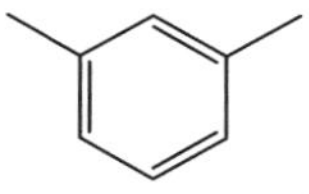

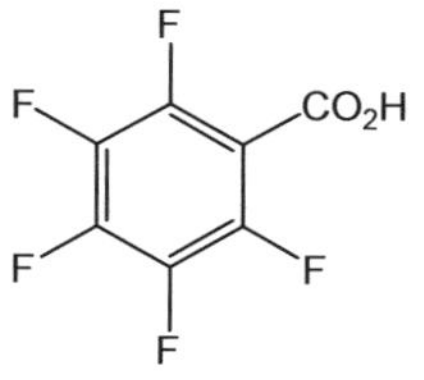

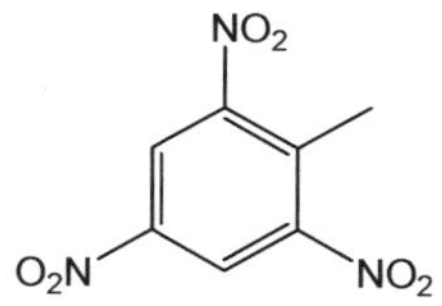

10. Draw each of the following compounds from their names.

     a.   *o*-nitroaniline

     b.   4-aminobenzaldehyde

     c.   *para*-hydroxybenzoic acid

     d.   *o*-methyltoluene (also called *ortho*-xylene or *o*-xylene)

     e.   *meta*-xylene

     f.   *p*-xylene

     g.   *m*-propoxynitrobenzene

     h.   2,3-dibromophenol

     i.   2,4,5-trimethylaniline

     j.   2,6-dimethoxybenzaldehyde

Notes

Nomenclature Worksheet 4

NAMING CARBONYL COMPOUNDS

Model 1: Aldehyde Nomenclature and Ketone Nomenclature

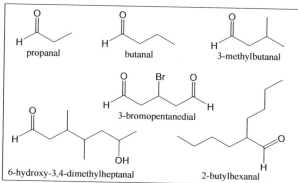

Aldehydes

Ketones

Questions (to do in or out of class)

1. Aldehydes and ketones each contain a **carbonyl group**.

   a. Describe the feature(s) that all aldehydes in Model 1 have in common.

   b. Describe the feature(s) that all ketones in Model 1 have in common.

   c. Describe structural difference(s) between an aldehyde and a ketone.

2. To name an aldehyde or ketone you start with the corresponding alkane parent name, remove the final "e" (for example "butane" becomes "butan"), and add a suffix.

   a. What suffix (ending) is added to complete the name of an aldehyde?

   b. What suffix (ending) is added to complete the name of a ketone?

3. Cite an example structure from Model 1 that demonstrates each of the following rules:

   a. The parent chain of an aldehyde must contain the carbonyl carbon.

   b. The parent chain of a ketone must contain the carbonyl carbon.

   c. You always number the parent chain of an aldehyde starting from the carbonyl carbon (even when there is an OH group present). *Note: By convention the "1" is understood and therefore omitted from the name (e.g., butanal, not 1-butanal).*

   d. The carbonyl C of a ketone gets numbering priority over other functional groups, even OH.

4.   Name each of the following compounds.

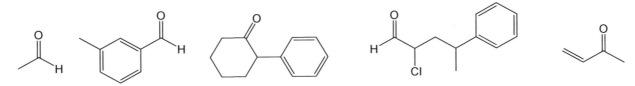

5.   Draw each of the following compounds from their names.

a. 2,2-dibromopentanal   b. 5-methylcyclopent-2-enone   c. 2-(cyclohex-2-enyl)cyclohexanone

Model 2: Carboxylic Acid Nomenclature

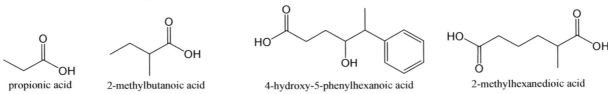

propionic acid   2-methylbutanoic acid   4-hydroxy-5-phenylhexanoic acid   2-methylhexanedioic acid

Questions (to do in or out of class)

6.   (*As with aldehydes and ketones*) To name a carboxylic acid you start with the corresponding alkane parent name, remove the final "e" (e.g., "butane" becomes "butan"), and add a suffix.

   a.   What suffix (ending) is used to name a carboxylic acid?

   b.   What word comes after the parent+suffix to complete the name of a carboxylic acid?

7.   Name each of the following compounds.

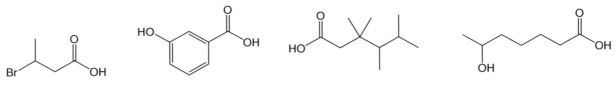

8.   Draw each of the following compounds from their names.

a. ethanoic acid   b. 3,3-dimethylbutanoic acid   c. 6-cyclohexylhexanoic acid

---

Memorization Task NW3.2: Common Names of Simple Carbonyl Compounds

Certain molecules are referred to so often by their common names that you should memorize them:

formaldehyde
(methanal)

acetaldehyde
(ethanal)

acetone
(2-propanone)

formic acid
(methanoic acid)

acetic acid
(ethanoic acid)
main component of vinegar

The prefix "**acet–**" is often used to name a two-carbon chain as in **acetic acid** and **acetaldehyde** (**acetone** is an exception to this trend), and the prefix "**form–**" is often used to name a one-carbon chain as in **formic acid** and **formaldehyde**.

Model 3: Ester Nomenclature

methyl formate
(methyl methanoate)

ethyl acetate
(ethyl ethanoate)

butyl propanoate

tert-butyl 2-methylpentanoate

2-chloroethyl acetate
(2-chloroethyl ethanoate)

Questions (to do in or out of class)

9.   Esters are one of class of "**carboxylic acid derivatives**" we will study. Esters are called this because an ester can be "derived from" (made from) a carboxylic acid. Based on the examples in Model 3, describe the structural similarities and differences between esters and carboxylic acids.

10.   (E) Each ester name consists of two words. What does the <u>first word</u> tell you about the structure?

11.   The <u>second word</u> in an ester name is generated from the corresponding carboxylic acid name.

   a.   What is removed from the carboxylic acid name?

   b.   What <u>three</u> letters replace "ic acid" to complete the ester name?

12.   Name each of the following compounds.

13. Draw each of the following compounds from their names.

    a. methyl acetate   b. phenyl 3-bromobutanoate   c. 2-bromopropyl benzoate

Model 4: Secondary and Tertiary Amide Nomenclature

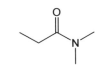
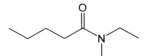
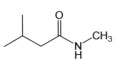
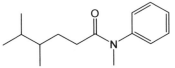

N,N-dimethylpropanamide   *N*-methyl-*N*-ethylpentanamide   *N*,3-dimethylbutanamide   *N*,4,5-trimethyl-*N*-phenylhexanamide

Questions (to do in or out of class)

14. Amides (like esters) can be made from carboxylic acids and so are considered another class of **carboxylic acid derivative**. Based on the examples in Model 4, describe the similarities and differences between the structures of <u>amides, esters, and carboxylic acids</u>.

15. Each amide name contains a parent name that tells you the number of carbons in the parent chain. Note that this <u>parent chain must contain the carbonyl group</u>. The parent name is derived by replacing "**oic acid**" in the corresponding carboxylic acid name with "**amide.**"

    a. Underline the parent name in each name in Model 4.

    b. Trace the corresponding parent chain on each structure in Model 4, and number the carbons, noting that the <u>carbonyl carbon always gets the number "1"</u>.

    c. Circle each alkyl group (if any) that branches off the parent chain.

    d. Circle each alkyl group (if any) that is attached directly to the nitrogen.

16. For an amide, the names of all the circled alkyl groups are listed before the parent name.

    a. What information does an italic capital *N* in front of an alkyl group (e.g., *N*-methyl) convey?

    b. What information does a number in front of an alkyl group (e.g., 3-methyl) convey?

17. Three of the amides in Model 4 are called **tertiary amides** because there are three C's bound to N.

    a. Which amide in Model 4 is a **secondary amide**?

    b. Draw <u>and name</u> an example of a **primary amide**.

Model 5: Primary Amides

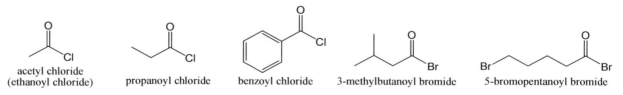

formamide (methanamide) acetamide (ethanamide) 3-methylbutanamide 5-bromo-5-methylheptanamide 3-methylpentanediamide

Questions (to do in or out of class)

18. Explain why you will never find an italic capital N in the name of a <u>primary</u> amide.

19. Identify the two common names shown in Model 5, and write the carboxylic acid common names from which these amide common names were derived.

20. Name each of the following compounds.

21. Draw each of the following compounds from their names.

a. *N*-methylformamide   b. *m*-nitro-*N*-phenylbenzamide   c. 3-phenyl-*N*-propylpropanamide

Model 6: Acid Halide Nomenclature

acetyl chloride (ethanoyl chloride)   propanoyl chloride   benzoyl chloride   3-methylbutanoyl bromide   5-bromopentanoyl bromide

Questions (to do in or out of class)

22. Below each acid halide in Model 6, write the name of the corresponding carboxylic acid.

23. Based on the examples in Model 6, construct a set of rules for writing the name of an acid halide from the corresponding carboxylic acid name.

Model 7: Acid Anhydride Nomenclature

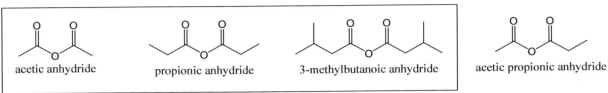

acetic anhydride   propionic anhydride   3-methylbutanoic anhydride   acetic propionic anhydride

Examples of symmetrical acid anhydrides

Questions (to do in or out of class)

24. Draw the structure of acetic acid below the structure of acetic anhydride.

25. Describe the similarities and differences between the structure of a <u>symmetrical</u> acid anhydride and the structure of the corresponding carboxylic acid.

26. Based on the examples in Model 7, construct a set of rules for writing the name of a symmetrical acid anhydride from the corresponding carboxylic acid name.

27. Draw the structure of benzoic anhydride.

28. Based on the example in Model 7, construct a set of rules for writing the name of an <u>asymmetric</u> acid anhydride from the corresponding carboxylic acid nam<u>es</u>.

29. Draw the structure of propionic benzoic anhydride.

Model 8: Naming the Conjugate Base of a Carboxylic Acid

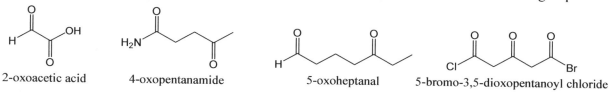

formate   acetate   propionate   2-methylbutanoate   4-hydroxy-5-phenylhexanoate

Questions (to do in or out of class)

30.   Based on the examples in Model 8, construct a set of rules for writing the name of the conjugate base of a carboxylic acid from the corresponding carboxylic acid name.

31.   For a given carboxylic acid (e.g., acetic acid), the name of its conjugate base (i.e., acetate) is found within the name of one of its derivatives. Draw the structure, and <u>write the name</u> of a derivative of acetic acid (found in this ChemActivity) that contains the word "**acetate.**"

32.   Draw the structures of **methyl benzoate** and **benzoate**, respectively. *Warning: Do not confuse the name of an ester with the name of a carboxylate (conjugate base of a carboxylic acid)!*

Model 9: "oxo" Nomenclature for Multi-Carbonyl Compounds

In a structure with different carbonyl functional groups, a carbonyl can be named as an **oxo** group.

2-oxoacetic acid   4-oxopentanamide   5-oxoheptanal   5-bromo-3,5-dioxopentanoyl chloride

Questions (to do in or out of class)

33.   Draw each of the following compounds from their names.

a.   2-oxobutanoic acid   b.   3,4-dioxopentanoyl chloride   c.   phenyl 3-chloro-3-oxopropanoate

Confirm Your Understanding Questions (to do at home)

34. Which names in Model 3 are common names, and which are IUPAC names? (The common names shown are used much more frequently than the IUPAC names for these compounds.)

35. Complete the analogy: ester is to alcohol as amide is to _____.

36. Draw the structure that corresponds to each of the following names.

    a. butanal

    b. 2,2-dimethylhexan-3-one

    c. 6-methyl-4-propylheptan-3-one

    d. 2-(2,2-dimethylbutyl)heptanal

    e. 2,6-difluorobenzoic acid

    f. 5,5-diphenylpentanoic acid

    g. 3-chloropentanoyl chloride

    h. cyclobutanone

    i. methyl acetate (or methyl ethanoate)

    j. 4-ethyl-2,3-dimethylhexanamide

    k. acetamide (or ethanamide)

    l. *N*-isopropyl-*N*-propylhexanamide

    m. 3-methylbutan-2-yl propionate

    n. benzyl benzoate

    o. 2,4,6-trimethylbenzyl acetate

    p. 2-phenylacetic anhydride

    q. phenylacetic propionic anhydride

    r. methyl 5-bromo-5-methylheptanoate

    s. disodium 3-methylpentanedioate (a dianion with two Na^+ counterions)

    t. potassium 3-oxohexanoate (an anion with a K^+ counterion)

    u. *tert*-butyl 2-methyl-4-oxobutanoate

    v. pentane-2,4-dione

    w. isopropyl 3-chloro-3-oxopropanoate

    x. *N*-isopropyl-*N*-methyl-2-oxobutanamide

    y. methyl 2-amino-2-oxoacetate

    z. *N,N*,2,2,3,3-hexamethylbutanamide

Notes

Notes